Wheels of Fortune

A couple's wondrous 14,000 mile bicycle ride around Canada and the United States

Chris Fieldsend

with Ties Benguedda

Copyright © 2021 Chris Fieldsend

The moral right of the author has been asserted.

All rights reserved. No part of this publication may be reproduced, distributed, or transmitted in any form or by any means, including photocopying, recording, or other electronic or mechanical methods, without prior written permission from the author, except in the case of brief quotations embodied in critical reviews.

Editor: Ross Daliday
Cover design: E3D Studio London Ltd
All maps: E3D Studio London Ltd, Adobe Stock, Jonathan O'Keefe's Strava Multiple Ride Mapper
All photos are authors own, unless otherwise stated.

Front Cover image: Golden Gate Bridge, San Francisco Bay
Back cover images (clockwise from top left): North Algodones Dunes Wilderness Area, California; Lake Erie with Alastair (Photo: Iwona Stepien); Cadillac Mountain, Acadia National Park, Maine; Parliament of Canada, Ottawa; Big Sur, California; Belvédère Beaulieu de Saint-Simon, Quebec

All stories in this book are true, but some names have been changed.

This book has no association with, and was not sponsored by, any of the commercial entities mentioned.

ISBN: 978-1-80049-977-5
First published November 2021
Publisher: Independent Publishing Network
This paperback edition first published November 2021

"Nothing compares to the simple pleasure of riding a bike"
– John F Kennedy

CONTENTS

	Prologue: Jailbirds	1
1	What the hell's a Paragon Rocker?	5
2	Maketh the wielrenner	15
3	Le grand départ	24
4	Walloped by a tornado	35
5	Mountains and Mennonites	48
6	Doodlepuffy	62
7	Super Typhoon Songda	72
8	Big Sur and big cities	86
9	Supertramps	101
10	Don't mess with Texas	112
11	Commander-in-grief	126
12	See you later alligator	140
13	Oh Carolina	153
14	George Washington's a Mackem?	163
15	Live free or die	176
16	Looney tunes	186
17	Big potatoes	200
18	Completing the loop	212
19	Screeched-in	222
	Epilogue: Ants in the pants	236
	Acknowledgements	239
	World Bicycle Relief	242
	Almanac	245

PROLOGUE

JAILBIRDS

I never expected to spend a night of our bike tour in a Texas jail. The Lone Star State has a renegade reputation and I didn't anticipate staying in one of its incarceration facilities, much less voluntarily.

The day in question was hot and sticky: we had grown used to weather like this in the Southern states. The campground we had lined up was either closed or abandoned. It was hard to tell which. We kept pedalling, on the lookout for a cheap motel.

Wild camping was not an option as our route took us close to the Mexican border and we weren't keen to take our chances. We weren't worried about those trying to cross the border, but we were terrified of those stopping people trying to cross the border.

After a few more miles we came across a small fire station. We rapped on the door but nobody answered, so we chugged on. Eventually we came across a tiny border town. The sort of town you expect Tombstone to be like, as opposed to the tacky tourist destination it is now.

The first building we came across was the Sheriff's Department. We parked our bikes and gingerly approached the open door of the small ramshackle building. There was an audible southern drawl resonating from the doorway, but no one to be seen, and they clearly hadn't clocked us.

We poked our heads into the office and saw the sheriff reclining in his chair, gleefully showing off his handgun to his deputies. Eventually he saw us, and, in one smooth motion jumped out of his chair, tucked his weapon into the back of his slacks, and let out a raucous "howdy."

All three law enforcers promptly adjusted themselves and joined us outside in the winter sun. It was made clear that we should directly address the sheriff. We promptly obliged and explained our plight. The sheriff's response was a little surprising.

"Y'all have a tent?"
"Yes, we are fully self-sufficient but wary of camping somewhere we shouldn't" I replied
"Surface y'all after?"
"Excuse me?"
"Y'all wanna camp on grass, asphalt, sand… What y'all after?"
"Oh, we're not bothered at all. We're just worried about it getting dark soon so keen to find somewhere quickly."

The cocksure sheriff went on to explain that we could put our tent up outside his building if we didn't mind sleeping on hard ground. He quickly followed this by explaining there was some grizzly weather coming in that night so we should push our tent close to the adjacent building. We enthusiastically thanked him and asked what the building next door was used for.
"Dat'd be the jail, but don't y'all worry, it's bein' renovated and not in use" said the sheriff, as if it were the most normal sentence in the world.
Then followed the magic words "S'pose with dis nasty weather comin' tonight y'all could stay in there if y'all like?"
And this is how we ended up spending a night in jail.
The sheriff instructed his deputies to give us a tour, which they did while apologising for the numerous bullet holes that pot marked the border town jail. We didn't care. In fact, you can't begin to imagine our delight when we discovered there was a kitchenette, toilet, sink and electricity.
As the deputies left us to set up for the night, one of them paused and apologised again, this time for the fact that the external door didn't shut.
"Don't shut, never has", said the sheepish deputy.

Prologue

While that seemed like a bit of a design flaw for a jail, again, we didn't care. We had a pub wash in the sink, took the obligatory photo of us behind bars and charged all our electronic devices.

Eventually we trundled up to the local mom-and-pop shop to try to find some fresh food. We returned with our usual hackneyed stash of Chips Ahoy Birthday Frosting cookies and Lays chips (crisps to you and me). Along with an extra pack of Chips Ahoy for the sheriff and his deputies.

Our hosts gratefully accepted our token gift and then the sheriff regaled us with tales of border town Tom and Jerry. He explained that he had had a successful career earning a seven-figure salary (unlike Brits, Americans aren't shy telling you what they earn) in the Big Apple, but wanted to return to Texas "to catch 'em bad guys."

The bravado with which the sheriff stood there – thumbs tucked into his enormous Texas flag belt buckle, hat angled forward and cowboy boots splayed at precisely forty-five degrees – was priceless. He told us with delight how many "damn Mexicans he'd caught crossin' the Rio [Grande]."

His favourite story was how he'd singlehandedly prevented a bloodbath from standing on the "good side of the Rio eyeballin' a gang leader."

Finger on trigger, he'd stared the assailant out until his foe backed down and hightailed it out of his weapons' range.

Before we returned to jail to bed down for the night, the sheriff made us aware of the one condition of our free night with America's most badass sheriff.

"Only thing I ask is that y'all don't tell anyone about this" adding "Don't want loads of damn cyclists turnin' up for a free stay."

So that is why I start this book with this story. Texas is a big place and this way I can tell you about our decriminalised night in jail while respecting the sheriff's wishes.

***

As a Brit, I was born to be cynical. It's in my DNA. But after living in a tent for a year, with no more security than a bicycle lock and a shifty beard, I have a far more positive outlook on life. And I feel incredibly fortunate for our many encounters with generous strangers, like the Texas sheriff, that made this trip a whole lot more enjoyable.

I hope you enjoy this book. I hope it restores your faith in humanity somewhat. I hope it makes you less worried about stranger danger and more open to strange encounters. I hope it inspires you to take flight and experience new things. Even better if by bicycle.

CHAPTER ONE

WHAT THE HELL'S A PARAGON ROCKER?

Like all good stories, this one starts in the pub. Specifically, the craft beer emporium that is The Rake, in London's Borough Market. It's the summer of 2013 and my friend Ross is excitedly telling me about a nine-month overland adventure he had seen around South America. Ross is well travelled but he'd done nothing on that scale. I thought he may be getting ahead of himself.

I met Ross back in 2002 in Soho, London. We were out with mutual friends experiencing the 'sleazy glam-rock extravaganza' night that is Decadence, in the underground pit that was Gossips nightclub. Gossips is sadly long gone, but my friendship with Ross has endured, despite his disdain for my love of hair metal and AOR. Over the years we have found common ground in nerdy rock sub-genres, travelling far and wide to watch live performances from titans of rock like Iron Maiden and Manowar, as well as lesser-known bands like Three Inches of Blood and Lord Weird Slough Feg.

A few days after Ross first infected me with the travel bug, we found ourselves back in The Rake sampling more craft beers. We were preparing our minds and bodies to see the angular metal band Deafheaven play at Birthdays in Dalston, east London. But the topic of conversation was not about Deafheaven's latest LP or how to avoid our eardrums bleeding. It seemed Ross was serious about the overland adventure. I was dismissive.

In 2013 my life was steady. I had a parent-friendly, but not overly exciting 'career job'. A year later I would read a book called the *Escape Manifesto* and discover I was stuck on the 'Travelator', an invisible force that hoodwinks us into thinking the steady path is the only option. More of that later.

Deafheaven were phenomenal but had given me a throbbing headache. Birthdays is a tiny basement venue and the bass-heavy riffing had not been ideal preparation for the next days' slew of meetings about meetings. Last night's music was not the only thing ringing in my ears; I couldn't stop thinking about Ross' overland adventure. The idea gnawed away at me, not just for the rest of that day, but for the next year. And it made me think, 'if I was to jack it all in and do this, how would I do it?' South America sounded amazing, but it's not like I've ever had a craving for it. And I'd never heard of this overlanding malarkey.

***

My parents, Margaret and Mally, met when they were sixteen, bopping to Bay City Rollers at the glitzy Court School of Dancing in Hull. My dad joined the Royal Navy soon after and was promptly posted to Portsmouth, Hampshire. I came along shortly after that and two years later I was joined by a baby brother, Adam. Dad was posted again, this time to Plymouth, Devon, where I spent the remainder of my childhood.

Plymouth has long been a strategic naval port, priding itself on its association with legendary boules player, privateer and slave trader Sir Francis Drake. During WW2 Plymouth's physical and strategic position in the UK's military arsenal put a bullseye on the city and it was largely decimated. There used to be a museum in Plymouth called the Dome, which had an exhibit commemorating the city's role in the war. The Dome didn't last long and the building is now occupied by cafes.

One of Plymouth's greatest landmarks and point bearers is Charles Church, an impressive seventeenth century feat of architectural design and engineering. Sadly, like much of Plymouth's medieval masonry, Charles Church is now a relic of its former past having been subjected to two nights of relentless bombing in WW2. For the past eighty years Charles Church has been known by locals as the 'Bombed-Out Church', having been left completely unrestored as a memorial to the city's many victims of Europe's infighting. However, this respectful and sympathetic approach to architectural legacy was not a strategy that the local mandarins applied throughout the city.

What Plymouth lacks in manmade beauty it more than makes up for in its physical landscape. Plymouth is surrounded by stunning coastline, and is blessed with the landlubbers dream that is Dartmoor National Park. Perhaps as a consequence of this, growing up in Plymouth, everyone I knew was either into surfing or cycling. I was definitely a two-wheeled enthusiast.

My oldest friend Neil and I would spend any time we could cycling. I had a neon green Raleigh Lizard, and Neil was the envy of all with the suspension forked Raleigh Activator. We would ride to the National Trust's Saltram Park and bomb along the Plym Valley Trail to 'The Rock' on the edge of Dartmoor. If we were feeling keen, we would head off-road up one of the iconic tors, just so we could slalom our way back down the rock-strewn bridleways. Looking back, it's incredible how well these early 'mountain bikes' handled the terrain and the abuse we threw at them.

In 1997 I bought and restored a MK1 Raleigh Chopper with the enviable date stamp of February 1970. This made it the oldest known chopper registered with the niche fraternity the *International Raleigh Chopper Owners Club*. The bike was a wreck and I sourced all the deadstock, reconditioned and CNC machined replacement parts through fanzines. Eventually I saved up enough money from my job as a toilet cleaner at the same school I attended, to have the whole frame and forks chrome plated. My efforts were acknowledged by Plymouth's *Evening Herald* newspaper, and in 1998 I appeared on the front and back page of the weekend supplement, resplendent in a tweed suit and Noddy Holder sideburns.

I continued riding mountain bikes through my teens, briefly dabbling in proper off-road downhilling. This didn't last long as it soon exposed my lack of technical skill and crummy bike handling. After a predictable hiatus of all things sporty at university, I rediscovered my love of cycling through the fixed gear craze that swept London in the early 2000's.

***

In 2007 I succumbed to another idea that originated in the pub, agreeing to take part in the 120 mile British Cyclosportive. This was a unique event, allowing amateurs the opportunity to ride the first stage of the Tour de France from London to Canterbury, the day before legends like Alberto Contador, Robbie McEwen and Tom Boonen would do the same. At the time my only bike was a cheap *Specialized Langster* which I rode fixed.

I set myself a gruelling training schedule, riding my *Langster* in all weather and combining training rides to visit friends in the surrounding counties of Kent, Surrey and Sussex. My hardest test – and still the toughest ride I've ever done to this day – involved cycling 130 miles on the fixie from London to Nottingham. The route was relatively flat but I had an awkwardly packed rucksack, with a heavy Kryptonite Fahgettaboudit D-lock strapped to the outside, flapping around whacking me in the ribs. The actual British Cyclosportive was nothing in comparison to this.

I arrived in Nottingham in agony, but my cousins Tony and Maria greeted me with more food and beer than even I could consume after burning 10,000 calories. The next day I met Ross at Download Festival in Donnington Park for live performances from favourite bands including Mastodon and Iron Maiden. After a few beers, the legs stopped complaining, and I was able to enjoy Dave Murray's galloping guitar, content in the knowledge I was probably the only person who'd cycled there from London.

Nowadays I couldn't make this claim. Since 2013, the Heavy Metal Truants, formed by Rod Smallwood (Iron Maiden's long suffering manager) and Alexander Milas (Editor of *Metal Hammer* magazine), have ridden from London to Download Festival. Rod and Alexander have somehow coerced unwitting band members and fans to complete the undertaking on rickety bikes. In June 2021, due to the Covid-19 pandemic, the first virtual ride took place. Iron Maiden's energetic frontman Bruce Dickinson led the charge and helped the Heavy Metal Truants raise over £1,000,000 for charity.

Weeks after the Download Festival, I was hanging around in Greenwich Park early doors, waiting to be called for the start of the British Cyclosportive. It was a glorious July day. The route was mercifully flat, with only a few climbs counting towards the King of Mountain (KoM) polka dot jersey challenge for the pros the next day. After the fourth KoM climb, I was feeling confident, and I blasted the final fifteen miles at an average speed of twenty-two miles per hour. I eventually finished a respectable 1,988th place out of 5,000 non-fixie starters.

In 2009 my friend Andrew spotted a too good to be true deal on a *Trek Madone,* and we both bagged one. I was instantly hooked, and my love affair with road riding proper began. Yet another night in the pub led me to sign up to a charity ride from London to Amsterdam. This saw me ride 330 miles in less than forty-eight hours through Kent, France and Belgium, before battling the famously flat but windy terrain of the Netherlands. I didn't know it then, but at the same time that I was riding to Amsterdam, Ties (my now girlfriend) was plotting her escape from the same city.

***

Ties (pronounced 'niece' with a 'T') was born and raised in the Netherlands to Moroccan parents. After a misspent youth she ploughed through a myriad of jobs as a bartender, barber and piercer, diligently acquiring the funds to study mental health at university as a mature student. By 2012 Ties was teaching neuro-linguistic programming (NLP) to students and budding therapists, and was also working with young adults with autism, supporting their personal development with workshops and day trips. She was applying her study professionally, but her heart and mind were dreaming about travel, specifically sunny California.

We didn't know it at the time, but by 2013 we were both contemplating pastures new. Ties was undeniably ahead of me. In July 2013 Ties left everything behind – quit her job, gave up her apartment, gave away all her possessions – and jumped on a plane to San Francisco.

Back in the UK, I was still ruminating Ross' overlanding adventure. I was starting to convince myself that something like that was in order, but deep down knew that nine months aboard an eighteen-wheeler in South America wasn't for me. It was a trip to Bristol – where Ties and I now live – that was the catalyst.

I was visiting my friends James and Emma who are enthusiastic travellers themselves. Flicking through one of their coffee table travel books – one of those books that has a title like '100 trips to do before you die' – I came across the Pacific Coast Bicycle Route. I had heard of people hiring a convertible and driving the west coast of America, but cycling? I was intrigued. A little Googling took me to the Adventure Cyclist Association (ACA) website and I didn't leave it for hours, much to the chagrin of James and Emma.

The ACA has mapped out 50,000 miles of cycle-friendly routes in the US. 50,000 miles! I was sold. I spent days reading the descriptions and ogling the photos of the many different routes. The Pacific Coast was a gimme. Coastal scenery has always been my favourite landscape and the opportunity to visit the real locations of favourite films including *The Goonies*, *The Lost Boys* and *Into the Wild* was too good to be true. I ruled out the interior routes, those which would involve riding thousands of miles through the monotonous farmlands of the 'flyover states' for months on end. That gave me an idea; I could ride the perimeter of the US. Following ACA routes, it could be done.

I returned to work and mulled this over for a while. The more research I did, the more excited I became. The route was even feasible weatherwise. There was something missing from this route though. Something big that could add markedly to the trip and the whole experience. Canada.

I have close friends in Toronto – Greg and Michelle – it would be a shame not to see them and it would be a shame not to see more of the country. I had visited Greg and Michelle before but only explored Toronto and Georgian Bay. The latter had blown me away. It took days for Greg and I to road trip round Georgian Bay, which is remarkably just one 'bit' of Lake Huron, the fourth largest lake in the world. I wanted to explore more of the Canadian landscape.

The route was coming together. Toronto to Vancouver, down to San Diego, east to Florida and back to Toronto hugging the Atlantic coast. But there was so much more to think about. I didn't know a great deal about touring bikes and equipment, and then there was all the practical stuff to think about like my job, the flat and how I'd pay for the trip. This is when I read the aforementioned *Escape Manifesto*.

The premise of the book is that two guys with high powered, well paid jobs realised one day that while they were excelling professionally, there must be more to life. The authors debunk the myth of the career job as the safe option and highlight the absurdity of the usual school to university to work to mortgage normalcy, which they call the 'Travelator'. I was definitely on the 'Travelator'.

I had for some time been considering jumping ship and trying something new, but always fell back on the comfort of the mythical career job. The *Escape Manifesto* gave me the shove I needed. Thus, I had a plan; quit the job, become an independent consultant and do as much consulting as necessary to build a travel fund.

In May 2014 I put the plan into action, registering my company (the imaginatively titled *Fieldsend Consultants*), applying for a business bank account (that ended up taking six months) and handing in my notice (the easy bit). I was renting so I figured I would just end the tenancy agreement. And that was that. It was as simple as the *Escape Manifesto* chaps made out.

Now all the practical stuff was sorted I could start the fun task of researching bikes and equipment. I also had to address some mild safety concerns. Like would I be held up at gunpoint in the US? And would a bear eat me in Canada? Having recently qualified as a 'Risk Practitioner' I decided that while the impact of any of these things was at best annoying, the probability was highly unlikely and my ability to mitigate even smaller. There wasn't exactly much I could do about America's love of firearms, and while bear sightings are indeed common in Canada, the number of attacks is tiny. Having done a less than thorough risk assessment I decided to fall back on a tried and tested approach: I just wouldn't think about it.

***

In April 2014 I attended the annual Bespoked handmade bicycle show at the Olympic Park Velodrome. I was window shopping framebuilders for the touring bike I'd use for my North America adventure. Top of my list was a local Hackney Wick based framebuilder, called Tom Donhou, whose metalwork I'd been following for a few years online and in books celebrating artisan framebuilders. It also helped that Tom's studio was a ten-minute walk from the flat I was renting on the River Lea.

In February 2015 I took the plunge and made the short walk alongside the frozen river to Tom's studio. Tom was keen on the commission but warned me there was a nine-month waiting list. That didn't deter me. It had to be a *Donhou*.

I thought I'd researched every component, nut and bolt to the nth degree, but Tom asked some surprising and intriguing questions, including:

What the hell's a Paragon Rocker?

- 'If you want a Rohloff [traditional internal geared hub] I assume you want a belt [instead of a chain]?'
- 'Why not get a Pinion [modern internal geared hub] rather than a Rohloff?'
- 'Are we going for a bushwhacking klunker or Euro tourer vibe?'
- 'Are Paragon Rocker dropouts okay?'

I didn't know the answer to any of these questions, which both excited and worried me. I was eager to find out what all these exotic sounding terms were and expand my cycling vocabulary. I was also about to commit thousands of pounds to a dream bike build, and realised I was way out of my depth. I came clean with Tom, and he was both very friendly, and very helpful. Incidentally, I opted for a belt, didn't go with the Pinion, chose Euro tourer vibe and elected for Paragon Rocker dropouts. I still don't know what Paragon Rocker dropouts are.

Meanwhile, whilst I was ruminating about head angles and fork offsets, 5,000 miles away Ties was sunning herself in Playa del Carmen, Mexico. The three-month trip to San Diego had gone rather well, and Ties decided she didn't want to go back to the Netherlands after all. Ties had met some fellow travellers in California who gave her all the insider knowledge of how to sustain the travellers lifestyle long term. She didn't need convincing. Armed with her many life skills, Ties had successfully travelled through much of the US and Central America, before settling in 'Playa' where she met Bert.

Bert owned several properties in Playa, and Ties had come to a mutual arrangement whereby she managed his Airbnb listings in return for 'mate's-rates' on one of his beachside villas. Ties paid her way by offering life coaching and writing articles for an English language newspaper. She set up an entrepreneurs network via the Meetup service, where she made friends and hawked her services.

Back in London my new life as an independent consultant was also going well, and I decided to celebrate this with a holiday. I was determined to catch some sun and wanted to go somewhere that guaranteed topnotch snorkelling. After exhausting every Google search imaginable, I was hopelessly undecided where to go.

An overnight work trip to Crewe, Cheshire, solved my conundrum. Accompanying me on this trip was my colleague and friend Sophie. Sophie knew I needed a break, and was adamant that she knew where I should go. Sophie said Mexico was lots of fun as a single traveller, was sunny and had topnotch snorkelling. Sophie was very convincing and it didn't take long to persuade me that Mexico was where I'd take my break.

CHAPTER TWO

MAKETH THE WIELRENNER

Cancun was my first experience of an all-inclusive resort and it wasn't my cup of tea. I shunned most of the all-inclusive trappings, favouring instead to spend my time taking day trips to a bird watching island and the famous Mayan ruins *Chichen Itza*. I also ventured out of the hotel to ride a BOB, a one-person submarine / underwater scooter thing. It was a lot of fun pretending to be Jacques Cousteau for the day.

After a few days in the picturesque town of Puerto Morelos, I wound up in Playa del Carmen. Originally, I had booked a hotel but after a few more sessions in The Rake, Ross had convinced me to switch to a hostel, claiming I'd meet loads of interesting people. I was sceptical but took the plunge and found a hostel with a private room. Ross was mildly irritated that I didn't fully embrace the dormitory experience but happy that I was dipping a toe in the hostel world.

I begrudgingly joined in the 'fun' evening activities laid on by the millennial hostel workers. Days were spent exploring local beaches and cenotes on *colectivos* – mini vans used as communal taxis – with two young travellers Francois and Aiden. Cenotes are subterranean freshwater wells where tourists and locals alike congregate to swim, seek solace from the sun, and take advantage of the free pedicures from the gregarious fish.

On my penultimate night at the hostel the staff laid on a cinema night. I noticed a gorgeous young lady decked out in immaculately assembled 1950's attire and emblazoned with colourful tattoos. She was trying to move a heavy bench on her own, and I leapt up to help.

"Can I help you with that?" I asked, rather rhetorically.

"NO THANKS, I'VE GOT THIS" came the terse reply.

I ignored the scalding response and grabbed the bench, awkwardly shuffling it towards me. We easily struck up a conversation and then Ties said, "I'll buy you drinks all night long if you can guess where I'm from."

This is Ties' favourite hustle. It works, and it has afforded her many free nights out. Ties is five foot nothing and speaks English fluently with an American twang. Dutch people are not short. And don't generally have long, curly dark brown locks. No one ever guesses that Ties is Dutch-Moroccan. If you're ever approached by someone in a bar matching this description, then hey, this book has paid for itself.

Ties had scalped another traveller into buying her drinks and broke the ice at the same time. She told me that she lived in a beachside villa with a tree growing in the middle and a resident snake.

"Of course you do" I said.

"I DO, I can show you if you don't believe me" instantly responded Ties.

We chatted for hours. Ties told me about the two years she'd spent travelling around the US and Central America. I told Ties about my plans to cycle around North America.

"That sounds awesome, good luck with THAT!" came the instant reply.

We left the hostel to check out Ties' serpent lair. After only a few paces Ties stopped abruptly.

"Can you ride a bike?", she asked.

I looked at her scornfully and reminded her about my plans to circumnavigate a continent by bicycle. Ties looked bashful, but still managed a matter of fact "Oh yeah."

Ties explained that she knew the hostel owner and hangs out there to partly fund her nomadic lifestyle by cutting travellers' hair. Within seconds of chatting to the owner, Ties reappeared with two rusty beach cruisers. We rode along Playa's Fifth Avenue, dodging drunk tourists looking for tacky cocktails, fast love and mariachi bands.

Ties' villa was indeed amazing. Behind a huge gate there was a secluded garden oasis with hummingbirds zipping between wild orchids and lizards scampering around rickety fences. Ties wasn't joking about the tree growing in the middle of the house and had underplayed the size of it. It was huge, so big the house must have been built around it, giving it a cool treehouse feel. The resident serpent was nowhere to be seen, but there were slithers of snakeskin in the tree that suggested Ties may not have been telling porkies after all.

I spent the night at Ties' and the next day there was an almighty monsoon. Ties left the room and minutes later burst back in manically with an open cutthroat razor. All of a sudden, I had flashbacks from the worst scenes of the trashy film *Hostel*. I knew I should have stayed in a hotel. I cursed Ross. This was his fault. Alas, Ties just wondered if I wanted a shave. Ah yes, I remembered, Ties is a trained barber and this is one of her hustles.

"Sure", I said, nervously.

After a few days together exploring the ruins at Tulum, and the wonderful Sian Ka'an Biosphere Reserve, we said our goodbyes and I flew back to the UK. We stayed in touch every day, and it was clear this wasn't a holiday romance. Ties had mentioned that it was her thirtieth soon and she didn't have any plans to see in the big day. I decided to take the bull by the horns and flew back to Mexico six weeks later for Ties' birthday.

Ties thought I was joking at first, but when she realised I was serious she was over the moon. We spent a few nights in Playa del Carmen, and then hopped on a local bus for the five-hour journey to Chetumal, a small city on the border of Belize. From Chetumal, we had a fun boat ride to Ambergris Caye, which we explored in a golf cart, discovering hidden beaches and dining in palapas. We also visited the idyllic island of Caye Caulker, where we snorkelled with manatees, nurse sharks and mesmerising spotted eagle rays.

It was a superb trip, and a few months later, Ties gave up her nomadic lifestyle and beachside villa to live with yours truly in Hackney Wick, east London. I softened the blow of moving to the UK in the autumn by preparing one of Ties' favourite dishes, gumbo – a hearty soup from the Southern US states – and then broke it to Ties that England were playing Wales in the Rugby World Cup that night. And I needed to watch it. I don't think this was how Ties anticipated spending her first night in the UK, but she did her best to enjoy it. I made it interesting for Ties by challenging her to guess the age of the rugby players. Ties was almost always two decades out; this is a game we have continued to play to this day. England lost in a nail-biting game. Ties was oblivious. She was still in shock that the beasts rolling around the grass were younger than her, some by ten years.

After a week, I felt I should address the elephant in the room.

"Remember that bike trip around North America I told you about?" I asked Ties over morning coffee, "Well, I'm serious about it and you should join me."

Ties' face contorted between giddy excitement and utter puzzlement.

"But I've never even ridden a bike with gears, or up a hill", Ties replied eventually.

I had to admit this made things challenging. I quickly thought about the route. I imagined the Rocky Mountains might be a trifle testing on a fully loaded bike if you'd never ridden up anything steeper than a canal bridge. Let alone never used gears before.

***

The Dutch are well known as a cycling nation. With 22.5 million bikes for 17 million people, no country on Earth comes close to the Netherlands' bikes per capita statistics. The Netherlands is also well known for being pancake-flat, with swathes of the country built on land reclaimed from the sea. What is less well known is there are two words for *cyclist* in the Netherlands. Most Dutch cyclists are referred to as *fietser (bicyclist)*, instantly recognisable from their leisurely upright position on the eponymous *opelfiets (granny bike)*. Then you have the *wielrenner (wheel runner)*, the lycra-clad, competitive whippets who adorn lightweight rigs with drop bars. Ties is a *fietser*.

"You'll be fine" I said, "I'll train you."

After another sip of her coffee, Ties placed her faith in me and agreed to join me. It was October 2015. I had planned to start the trip the following summer, giving me about eight months to turn Ties into a *wielrenner*. In my head it was easy. I simply needed to eke up the mileage each month, adding a little weight to the bike and gradually introduce Ties to the concept of cycling uphill.

Before Ties moved to the UK, I picked her up an old step-through *Raleigh Caprice* from eBay for £30. She loved it. Baby blue with a wicker basket and a Sturmey Archer three-speed hub. It was perfect for cruising around London, exploring the canals and city streets. It wasn't ideal for cycling thousands of miles around North America. Or even for training for such an endeavour. Regardless, Ties was determined and we took the *Caprice* out to Essex for a few autumnal rides. Enthusiasm understandably waned when winter crept in and we started to up the mileage.

At the end of October 2015 a magical email pinged in my inbox; I was next on Tom Donhou's framebuilding list. Tom had more questions. Exciting stuff like colour. I wanted a utilitarian look that didn't draw attention to the bike or me.

"Can it be painted in drab olive, like a *Willys Jeep*" I asked.

Tom approved of the idea and we were in business.

All this excitement made me realise we needed to find a *Caprice* replacement for Ties. An important factor in choosing a bike was Ties' diminutive stature. This narrowed things down, and I quickly discovered that British brand Genesis did an excellent touring bike (the *Tour de Fer 20*) in extra small. Weeks later, we picked up the Genesis from the local bike shop, replacing only the saddle for a touring-friendly *Brooks B17S*.

We booked our flights to Toronto soon after and started the lengthy and convoluted process of applying for a US visa.

Toronto was the planned start and end point of the trip for many reasons. Our friends Adrian and Pornipha's wedding was in late June 2016, so we couldn't leave before then, but we did want to maximise summer months. Starting in Toronto also meant we could spend some time with Greg and Michelle before we set off. Weatherwise, an early July Toronto start meant we would have summer in the most northern stretch of the trip, autumn down the Pacific coast, winter in the deserts of the Southern US states and spring on the Atlantic coast. On paper this should have been perfect. On paper...

***

Over the next few months, we gradually acquired our kit and, as planned, gradually upped the mileage and weight on rides. To make things interesting we combined the training with visits to friends and family, adding a social element and quelling Ties' thirst to explore new areas. Ties conquered some almighty climbs in and around London. We created a local hillclimb circuit, encompassing Swains Lane, Muswell Hill, Archway and Alexandra Palace. London is generally flat but these are nasty hills by anyone's standard, let alone a *fietser* on a fully loaded steel tank.

Ties notched up some eminent cycling climbs, including Box Hill and Ditchling Beacon, and further afield we did some gruelling rides in the Peak District, Forest of Bowland and Dartmoor. One memorable ride en route to the UK Cycling Touring Festival saw us climb 5,600ft; more than any day on our year cycling around North America.

Excerpts from Ties' Training Diary

30/04/16, Dartmoor: Chris was mentally prepping me for the extreme amount of hills. I tried to not let it discourage me but I did feel a bit scared. The landscapes on Dartmoor were so beautiful. Little creeks, pine trees, and golden spiky bushes everywhere. We saw so many wild horses, lambs and sheep. I absolutely loved it. But it was extremely hilly. No flat surface at all! There was a sportive going on: a lot of cyclists with numbers on their bikes passing us and giving us thumbs up for riding Dartmoor fully loaded. That did encourage me to ace those hills, including a four-mile climb. Chris' parents came out to meet us at Princetown (the highest point on Dartmoor) with a fantastic picnic. After lunch I was tired and wanted to turn around earlier than planned. I was all dramatic in my head, doubting whether I could do the big trip. This really taught me how important it is for me to be in a positive and optimistic state of mind.

26/05/16, Peak District: I think I passed the test! We cycled over nine hours, sixty-two miles of non-stop massive hills through the peak district (5,600ft elevation). Fully loaded at fifty-five-kilograms. Waaauw what a ride that was. One hill right after the other, not a single flat road. That was incredibly intense but unbelievably beautiful. I was feeling fine for most of it, but at the fifty-six-mile point I had had enough. My body started to hurt and I was mentally tired. I felt a tantrum coming up but Chris wisely ignored that. I wiped away my silent tears and pushed on. I slept like a baby that night, pleased with that victory and bracing myself for another day similar to the one that just ended.

Ties was fast making the transition to a *wielrenner*. The US government gave us their blessing for the trip, adorning us with six-month visas, and we even found travel insurance that would cover a year of cycle touring. Everything was going well. Something had to give.

On 20th February 2016 David Cameron (then UK Prime Minister) finally relented to catcalling from dubious right-wing politicians like Nigel Farage, by announcing the date for the EU referendum: 23rd June 2016. The landmark event dominated the news between February and June 2016. It dominated everything. Speculation was rife, but general consensus was this was a shoo-in for Cameron and his young Cabinet. He obviously thought so as well, as he neglected to do almost any campaigning. And we all know how that ended.

I figured our budget would mean we'd camp most of the time but spend a few nights a week in motels, chalk up microbreweries in each town and visit famous tourist landmarks. Friends and family were curious, regularly asking if I was stockpiling dollars in the event of a Vote Leave victory.

"Nah, we won't be leaving the EU, I have faith in my fellow Brits" I scoffed.

I had too much faith.

On 23rd June 2016 I did my last day of consulting for a year, jumped on the train back to London and strode into the voting booth. I proudly emerged with my 'I voted' sticker, congratulated myself with a few beers and trundled off to bed. The next day I rose early to collect the van I'd rented to move out of the flat that same day. Results were in, Nigel Farage had won and Brexit was a done deal.

The young guy who served me at the van rental place looked as forlorn as I did. As he showed me around the vehicle, he seamlessly switched between explaining how every knob and dial worked and lamenting the results of the referendum. I was checking out the cavernous payload at the back when I heard a loud 'SHIIIIIIIIIT' from the cabin. The rental chap had turned the ignition on to check the mileage, the radio came on automatically and David Cameron was announcing his resignation in real time.

The pound took an almighty nosedive, the country was rudderless, and the UK was a worldwide joke. Populist politics was set to make an international comeback and Farage was the smuggest man alive. This raised a number of questions. How would we afford the trip? Would Ties be allowed back into the UK after a year in North America? Where would we live if she wasn't allowed back in? What sort of reception would I receive in the *Land of the Free* after the Brexit result?

Many of these concerns became real issues, but I needn't

have worried about the last question. Four days into our first stint in the US, on Ties' thirty-first birthday, the Republicans nominated *Home Alone 2* and *Zoolander* star Donald Trump as their nominee to succeed the transformative Barack Obama as president. Populist politics was indeed back, and the US was making a play to become its global proponent. They welcomed me with open arms.

CHAPTER THREE

LE GRAND DÉPART

7th – 15th July 2016
Distance ridden this chapter: 432 miles / 695 kilometres

Adrian and Pornipha's wedding was awesome, and fortuitously for us, it was in Brighton: only a short taxi ride from Gatwick airport, from where our adventure would begin. We spent our last few days partying, gave our wedding outfits to friends James and Emma for safekeeping and jumped into a taxi. We were underway!

Anyone who's ever taken a bike bag to an airport knows how awkward it is to manhandle. Bike bags are tall, long, narrow and cumbersome. They have wheels on the bottom but unless you pack them perfectly, they end up hopelessly lopsided and take on a life of their own. Our bike bags weighed over thirty-kilograms each. We also each had a twenty-kilogram-plus body bag-sized duffel bag, which we'd bundled all our panniers into. The looks we received from unseasoned holidaymakers were extraordinary. Children were the most brazen, tugging on their parents' arms, while openly pointing and asking their mums and dads what we were lugging around the terminal.

Hours later we were in Toronto. This was it. I'd spent almost three years thinking about this trip and now we were on Canadian soil. Ties was beyond excited, 'doing an Eddie Murphy'; a phrase she coined based on the actor's ecstatic reaction to first exploring New York as Prince Akeem in the 1988 film *Coming to America*. From the first time she saw *Coming to America* Ties has felt an affinity with Prince Akeem. She understands the overwhelming excitement of exploring new places and had the same reaction on arriving in Toronto as she'd had starting her own 'three month' trip in California a few years earlier.

We wrestled our bikes onto the *Union Pearson Express*, the airport rail link which had only opened the previous year, in time for the 2015 Pan American Games. Our friend Greg met us at one of the calling points and somehow bundled all our stuff, and us, into his car.

I love Toronto. Since my first trip there in 2005 I've been in awe of its parks, pubs, culture and laid-back vibe. Ties was equally impressed and was bowled over by the urban wildlife. Racoons, skunks and giant black squirrels are all common sights, considered pests even by locals. Ties couldn't understand this, delighted every time she spotted any of these Canadian critters.

We made sure the bikes were properly reassembled and then spent a few days cruising Queen Street West, exploring Toronto Island and window shopping in Kensington Market. Toronto is the most diverse city on Earth. Over half the population were born outside Canada and its residents speak over 140 languages. There's a Little Portugal, a Little India, a Little Korea, a Little Most Places. Ties was 'doing an Eddie Murphy' on every block.

Eventually we had to leave. On 7th July 2016, in the midst of a baking heatwave, we said goodbye to Greg and Michelle and started our adventure.

***

Before we left the UK I had mapped out the entire trip, all 12,000 miles (as I thought it would be). I knew we wouldn't follow the route exactly, but I wanted a rough idea of where we'd be and when. This was important for timing the seasons, arranging to stay with friends who live en-route and scheduling to meet people who wanted to fly out and visit us during the trip.

I mapped routes along roads that looked both tarmacked and quiet. It was also important that the route took us through areas where we could acquire regular provisions and find places to stay. This was a tricky balance. For the US legs, I more or less followed the interactive map on the ACA website with detours where we wanted to see certain sights or people. Once I had the routes mapped out, I transferred them to my Garmin, which navigated us with an arrow that looks like Pac-Man moving around a map.

As well as mapping out the route I spent months before we left poring over camping websites, cycle touring forums and blogs deciding what tent, sleeping bag, air mattress and stove we'd take. Weight is obviously crucial with kit but as we were fully loaded – each carrying six bags on our bikes – performance and durability were more important. I discovered that camping stoves are a minefield and probably the most hotly debated bit of cycle touring kit. I read numerous blogs, most of which were highly biased towards one fuel type or another, extolling their own idiosyncratic virtues.

I wasn't completely green to this, having spent twelve years in the Scouts. I'd always used a stove with a spirit burner, but had a traumatic experience with one on a three-day mountain biking trip over Dartmoor with Neil when we were fifteen. I was boiling water when I heard a rumbling that didn't sound right.

"Stand back, I've got this" I said to Neil, confident I knew what to do from all my Scouting experience.

I took the water off the stand and as I did that the spirit burner exploded, spraying my bare arm in hot fuel. My arm was on fire. It hurt. And it caused some olfactory anxiety. I pegged it to the nearest stream and jumped in. I had plump blisters all over my arm for months and impressive scars for the next decade. I decided against a stove with a spirit burner.

I contemplated alcohol-based stoves, which draw fuel from a refillable cannister, but ruled this out because of the mess and frequency with which they need topping up. This left cartridge-based stoves, which generally use a butane/propane mix. Online opponents to these stoves frequently point out the problem with finding cartridges with the right connector, especially in less developed or remote areas. We were going to North America though, so I didn't see this as an issue.

***

Day one was an inevitable drudge through Toronto's industrial estates and retail parks. The highlight of the day was the ride through Holland Marsh wetlands and farms, where there were Dutch references everywhere we turned. We read on the plane to Toronto that the Dutch have an affinity for Canada because of the Canadian support during the dying stages of WW2. There is clearly still a strong connection, and we amused ourselves seeing who could spot the next Dutch name on mailboxes and business hoardings.

As we approached Barrie, a small city on Lake Simcoe, we found a Canadian Tire. There is no equivalent to Canadian Tire in the UK. Canadian Tire's own marketing describes it as 'Canada's Top Department Store' and this tagline neatly shows the difference between the UK and Canada. A UK department store involves running a gauntlet of perfume sprayers to search for a new suit, pillows or curtains. Canadian Tire is more like a camping, DIY, car and sports shop rolled into one. In short, 'Canada's Top Department Store' is more Go Outdoors than John Lewis, and definitely a cycle tourer's friend.

I strode in and found the stove cartridges easily. The staff were super friendly so I thought I'd push it and ask to use the toilet. The shop assistant looked me up and down quizzically in my lycra bib shorts and asked me to repeat myself.

"Can I use your toilet please" I said.

"I can show you *where* our toilets are" said the shop assistant.

"Great, thanks so much."

After a lengthy walk around the cavernous store the shop assistant led me to the bathroom section.

"Here you go" he said, pointing at a range of gleaming porcelain thrones.

I was confused. So was he. We both stood there, shuffling on our feet, me wondering where the door to the toilets was, the shop assistant wondering what a Brit on a bike wanted with new sanitaryware. I couldn't bear it any longer.

"Sorry, *where* are the toilets please" I said, uncharacteristically adding "I'm desperate."

After another uncomfortable shuffle, the shop assistant had a 'Eureka' moment, "OH, you want to use the restroom!"

Ties thought this was hilarious, mocking me for my ignorance of North American lingo as we rode the final few miles to our campground. I excitedly unpacked our stove and the Canadian Tire cartridge. However, something was up, it wouldn't attach to the stove. Ties mocked me for the second time that day; in the time it had taken her to erect the tent I was still wrestling with the stove. Try as I may, it would not attach. I had to concede I'd bought the wrong one.

Ties was hangry – an undesirable physical condition that causes irritation from a lack of food – and I was in the doghouse. After cycling fifty-nine miles through sweltering temperatures, we spent our first night eating the emergency nuts and energy bars we had intended keeping for, well, emergencies. Not a great start and I was lamenting my decision to bring the stove, wondering if we'd ever find the right cartridges and if all those online naysayers were right about the benefit of alcohol stoves after all.

***

Our goal on day two was Craigleith Provincial Park, fifty miles away. The route took us across flat farmlands and followed the Nottasawaga River to the white sands of freshwater Wasaga Beach. After admiring Georgian Bay for a bit, we jumped on Highway 26 and headed for Craigleith. We weren't sure if we should be cycling on Highway 26 but it was gloriously smooth and had a massive shoulder so we chugged on. After a few miles we spotted a lady excitedly flapping her arms up and down, beckoning us to stop. We were certain she was going to tell us off for riding on the road, so we sheepishly freewheeled over to apologise.

Rather than admonish us for being irresponsible, Barb, the lady who waved us down, wanted to invite us to a local town planning meeting about cycling and economic development. She said her house was a few miles past Craigleith, and we were welcome to stay the night with her and her husband Chuck. We gladly accepted Barb's invitation and followed her directions for the rest of the day. Barb and Chuck's house was easy to find: they'd made us a welcome banner and left it tied to Barb's bike at the end of their drive.

It was only day two of the trip and we were experiencing some famous Canadian hospitality already. Barb and Chuck had never been cycle touring but were keen members of the local cycling club and were fascinated about our trip. They introduced us to their amazing guest suite, where we'd spend the night, and then we all headed down to the Beaver Valley Community Centre for the local event.

The town council had flown an economic development expert over from Washington State and had a packed agenda to get through. Despite this, Barb interjected early on and announced our presence as guests of hers from the UK. The residents were eager to hear our thoughts – as cycle touring aficionados – on what they could do to improve the local infrastructure. This was slightly awkward, as we were only two days into our cycle touring 'careers' so very much novices ourselves. Also, the expert from Washington State seemed a little perturbed at being bumped down the agenda by a pair of cycling hobos. Regardless, we gave it our best shot, extolling the virtues of Canada's wide shoulders but highlighting the additional benefits of Dutch style fully segregated cycle lanes. Our insightful feedback was greeted with lots of nodding, beard stroking and a ripple of light applause.

The meeting was interesting and it was great to find out more about the local area and meet local people. A buffet was laid on in the car park, but as we were about to eat an almighty thunderstorm came in. Within seconds all the gazebos were either bent double or tumbling around the car park. After several tense minutes running around collecting flapping canvas and rescuing prawn vol-au-vents, the organisers wisely decided to relocate indoors. After a delicious feed, we headed to our host's friends' beachfront house for drinks, and waymarking tips from the local cycling club route master.

Early the next day, Barb made us a scrumptious breakfast, and Chuck drove us around the local area to try to find a stove cartridge with the right connector, to no avail. We thoroughly enjoyed our stay with Barb and Chuck, and we have remained in contact to this day. We set off from Barb and Chuck's brimming with enthusiasm for what the route offered and wondering who we might meet next.

Back on Highway 26 we eventually found the right stove cartridges at Home Hardware in Meaford. Phew, we could use the stove in North America after all! From Meaford, we rode along the scenic Georgian Bay shoreline to Wiarton. The lakeside town is home to the weather-forecasting groundhog *Wiarton Willie,* Canada's answer to *Punxsutawney Phil,* the animal star of the 1993 film *Groundhog Day*. We found a fantastic campground on the shore, close to the prominent *Wiarton Willie* statue, and as soon as we rolled onto our pitch, a couple from Guelph offered us dinner and a beer. We gratefully accepted, inhaling both before setting up camp for the night.

***

Our plan from Tobermory was to take the *Chi-Cheemaun* ferry to Manitoulin Island; the largest freshwater island in the world. *Chi-Cheemaun* means 'big canoe' in Ojibwe, the second most common language of Manitoulin Island, spoken by the First Nation habitants who make up over forty percent of its population.

Whilst waiting for the *Chi-Cheemaun* we met Iwona, Kasha and Irek who invited us to stay at their cabin on Manitoulin Island's Kagawong Lake. We also met a cycle tourer from Hamburg called Knut, who was heading to the Manitoulin Conservatory for Creation and Performance (MCCP) or the *Clown Farm* as it's more commonly known. The MCCP was established by a Toronto couple who thought that Manitoulin Island was the ideal location for clowning lessons, believing its natural landscape would stir creative spirits amongst their students; the island's name is an Anglicised version of *Manidoowaaling*, Ojibwe for 'cave of the spirit'. Knut certainly thought so and was thrilled to be combining his passion for cycle touring with his ambitions to learn the comedic artform. We rode with Knut to Providence Bay, where he honked an imagery horn at us as we peeled off to Kagawong Lake.

The last three miles to the cabin were along a challenging gravel road, and we needed to push our fifty-five-kilogram rigs for the first time on the trip. The trek was well worth it though. The A-frame cabin and outbuildings were perched right on the picturesque Kagawong Lake, with no neighbours except the excitable woodpeckers. The family explained how important it was to them to live sustainably and make a minimal impact on the physical landscape. This involved some practical compromises and their smallholding had no electricity or running water, and a homemade compost toilet. We embraced the good life by bathing in the lake and then listened to Kasha playing the ukulele, before passing out under the stars. It was easy to imagine how this setting would help Knut unleash his creative spirits.

We left Kagawong Lake on an extra hot day that seemed to offer no shade on any of the roads I'd chosen. We were making decent progress until we hit a steep hill past Whitefish Falls. On our entire trip Ties had three meltdowns, all weather-related. This was one of them. She was struggling with the combination of intense heat and lack of shade, and the ten percent gradient of that hill was the final straw. She dismounted her bike and pushed it down the verge. Ties was a wreck and determined that we finish the day there and then.

I took my phone out to look for nearby options, but all I could find was Widgawa Lodge & Outfitters. I kept looking. No joy. Ties was upset, ruined by the heat and the climb. 'Sod it, let's see how expensive Widgawa Lodge is' I said, to myself. To my utter delight Widgawa Lodge *was* a campground. I consoled Ties into cycling a little more and we pressed on.

The access road to Widgawa Lodge was giving off proper *Wrong Turn* vibes. The tarmac disappeared rapidly and was replaced by an uneven gravel track with haunting trees blocking out the sunshine. The sides of the road were scattered with abandoned vehicles growing foliage through their glassless window frames and there wasn't much sign of life. I could hear Ties' inner voice swearing at me and I was crossing all my fingers and toes that it was going to be open, let alone that we'd survive an encounter with *Saw-Tooth*. As we approached the end of the track, we emerged from the trees into bright sunshine and the pleasing aroma of woodburning stoves. We unpacked and set up camp, saying a thankful prayer to the Gods of Cycle Touring, and spent a peaceful and pleasant night amidst the sounds and smells of the Canadian forest.

In Canada it's important to leave your food outside the tent; a vital precaution against bears smelling your edibles and making an unwelcome night-time entrance. The owners at Widgawa Lodge said we could store our food panniers in one of the empty chalets that night. We did not expect the scene that awaited us the next morning. It looked like the chalet had been turned over by burglars. Our panniers weren't where we'd left them and there were remnants of food everywhere. On closer inspection, our panniers were covered in tiny paw prints. Racoons! They'd broken into the locked chalet, opened up our buckled panniers and helped themselves. Maybe they are pests after all, we thought. The first of many lessons to be learned about cycling in Canada.

***

The worst roads we experienced in Canada and the US were both called Highway 17. On day eight we started our stint on the Canadian version. It was horrible. Logging juggernauts blasted past us as we tried staying in the tiny shoulder, fighting the crosswinds that were trying to push us out into traffic and a messy demise.

We peeled off Highway 17 momentarily to stop in Thessalon for lunch, a town whose name sounds like it is derived from the ancient Greek city, but is thought to be a corruption of the First Nation term 'Neyashewun', meaning 'a point of land'. We grabbed the bench outside the town's library and heated up a can of Stagg Chili, much to the amusement of passing 'Thessalonians'. While tucking into our chili we spotted a Home Hardware and took the opportunity to pick up some thin rope. We decided to take better precautions against Canadian wildlife and had read that we could create a makeshift bear locker with a pannier and some rope.

Our second day on Highway 17 ended at an agreeable campground at Bruce Mines, on the north shore of Lake Huron. I decided to give the homemade bear locker a go. The trick was apparently to secure the pannier at least twelve foot off the ground but also hang it six feet from the branch, away from the mitts of bears and raccoons. I tied the rope around a stone and tried to throw it over a high branch to winch our food pannier up. Try as I might I couldn't do it. I failed miserably.

Our camping neighbour, Greg, thought this was hilarious. After entertaining himself watching a ridiculous Brit trying to do something quintessentially Canadian, Greg came over and offered to store our food pannier in his enormous RV. I gladly accepted but sheepishly explained that we normally make breakfast at 06:30.

"No problem, do you drink coffee" was Greg's instant response.

And sure enough, as I wandered back from the facilities just after 06:00 the next morning, Greg was sat outside his RV with our food pannier and a vat of piping hot coffee. The last day of our first leg in Canada had started wonderfully.

CHAPTER FOUR

WALLOPED BY A TORNADO

16th July – 20th August 2016
Distance ridden this chapter: 1,848 miles / 2,974 kilometres
Total distance ridden: 2,280 miles / 3,669 kilometres

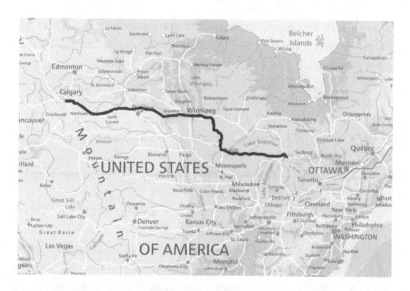

There are five massive lakes on the Canada-US border that are collectively known as the Great Lakes. Lake Superior, Michigan, Huron, Ontario and Erie all feature in the top thirteen biggest lakes in the world, as measured by surface size.

Lake Superior – as its name suggests – is particularly massive. In fact, it's the world's largest freshwater lake and only the Caspian Sea (which is brackish) is a bigger inland body of water. If you drained Lake Superior, the water from all the other Great Lakes combined wouldn't fill it, and the displaced water would flood the continents of North and South America to a depth of one foot. These are not just interesting facts, they are details that forced us to make the first tough decision about our route: do we go south or north of Lake Superior?

Distance-wise, there's almost no difference. The southern route is negligibly longer, but it's also much flatter so easier going overall. The northern route meant two weeks on the dreaded Highway 17. We'd read scary things about this section of our already least favourite road and we didn't fancy subjecting ourselves to a fortnight of misery. We elected to go south, dipping into the US and riding through Michigan, Wisconsin and Minnesota.

What we didn't know was that by entering the US at Sault Ste. Marie, it meant the clock started ticking on our US Visa. This is how the US Visa system works:

- Fill out a lengthy application form and book an appointment at the US Embassy
- Queue for hours outside the US Embassy, despite having a timed appointment
- Take time ticket and wait around for more hours for application interview
- Be interviewed and take time ticket for security interview
- Wait for more hours and then, after all that, spend thirty seconds having a token interview
- Wait at home for passport to arrive with visa attached

After all that, the actual length of your visa is then determined by the Border Patrol Agent on the day you enter the US. The Border Patrol Agent can authorise a stay of between two weeks and six months. So despite going through all the financial and security vetting, if the Border Patrol Agent is in a foul mood or doesn't like the cut of your jib, you're stuffed.

Back in London, at our US Embassy interview, they told us that if we enter the US at Sault Ste. Marie we would be given a visa stamp for the time needed to re-enter Canada and then we'd be given a new visa stamp when we re-enter the US at

Blaine, Washington. This turned out to be complete twaddle.

The brutish Border Patrol Agent at Michigan thankfully granted us six-month visas. However, as he took giant swigs from his gallon jug of protein shake, he assured us we'd misheard things at our US Embassy interview. He said that if we wanted to remain after the six months, which started immediately, we'd need an extension. He was also puzzled by our plan, insisting that it wasn't safe to cycle around North America and nigh on impossible to obtain a US visa extension in transit. We said we'd take our chances and he waved us through with our I-94 admission record, shaking his head emphatically.

We were exhausted after our stressful introduction to the US but we couldn't end the day at the border and we rode for an hour until we rolled up to Brimley State Park. There was a large sign clearly stating they were full. Undeterred, we approached the park rangers and asked if we could stay. They explained that every pitch was prebooked, but they could clearly sense our desperation. After a bit of wrangling, they explained that there was somewhere they could put us but it wasn't very pleasant and had no facilities. We followed their directions and were delighted to find a level field, far away from all the RVs and bickering children: it even had its own long drop (toilet).

"Can we stay for two nights please" we begged.

"You can stay for as long as you like, no one ever uses this" the park rangers confirmed.

***

After a slothful rest day at Brimley State Park, we rode over 150 miles through relentless heat on long, straight roads with no shade, ending up in a small town called Harvey. It was Ties' birthday and we decided to push the boat out, booking a room at Americas Best Value Inn. We sampled some beers at the local Chocolay River Brewing and then ambled over to the Dry Dock Bar and ordered the 'Fully Loaded Pizza' and 'Monster Burger', with extra everything. The waitress tried to reason with us but we assured her we'd eat it all. We polished it all off with aplomb, gaining kudos and high fives from the beguiled staff.

We zipped across Wisconsin in two days and entered Minnesota over the massive Bong Bridge. The imposing two-mile structure is named in honour of Major Richard Bong, the WW2 'Ace of Aces' who was born in Superior, Wisconsin. The Bong Bridge connects Superior to Duluth, Minnesota, where our route took us up the only hill we rode in the exceptionally flat US state.

By this point we were two weeks in and growing fond of community campgrounds. North America is big on camping. But 'camping' in North America tends to be an activity undertaken in luxurious gargantuan RVs. As such, most campgrounds tend to have concrete pitches and many don't even allow tents – or 'tenting' as they call it – and those that do allow tents are invariably expensive.

Then you have community campgrounds. These come in many forms but are nearly always quiet, grassy, with well-maintained but basic facilities, fresh water and an honesty payment system. Our first night in Minnesota was spent at the community campground in Floodwood, a tiny city that has given itself the title of *The Catfish Capital of the World* owing to its annual *Catfish Days* Festival. The glorious community campground was set in a small wood on the banks of the St. Louis River. It was completely empty and for $15 we had all the drinking water and power we could utilise, plus the added bonus of excellent showers.

McCarthy Beach State Park was next on our itinerary. It had recently been voted one of the seventeen best beaches in North America and we were excited to discover how a freshwater beach could attain such a lofty status. We were decidedly underwhelmed by the distinctly ordinary lakeside beach, but the campground was fantastic and we had the joy of falling asleep to the haunting sound of Timber Wolves howling at the moon. We woke up early the next day and cycled the first hour in pouring rain. We were drenched but the rain seemed to make us go quicker and the ninety miles we rode seemed easier than some much shorter days.

At International Falls we crossed back into Canada, with no more interrogation than a smile and a wave. Relieved at the friendliness of the Canadian officials we made our way to Caliper Lake Provincial Park, where we hoped to stay for the night. Caliper Lake was spectacular. We stayed at over 100 campgrounds on our trip and Caliper Lake was my favourite, by a mile. Our pitch was right on the lake, and we had a picturesque beach with a floating dock all to ourselves. The icing on the cake was the scurry of red squirrels who ate raw almonds from our hands and happily crawled all over our tent.

The ride from Caliper Lake to Rushing River was out of this world. There were huge rock formations by the roadside, ancient lakes and things rustling in the leaves the whole way. A few miles after Rushing River, we spotted something strange: we could see the familiar sight of panniers, but they were all on enormous bullhorn handlebars and the rider was standing up. The pilot of the strange contraption was Davide, a crazy French guy who explained that he had 'ridden' across Canada, from Vancouver, on his Traczer scooter. It would be more accurate to say that Davide had 'pushed' his way across Canada, as his scooter worked by placing one foot on the central platform and pressing the other foot on the ground to propel himself forward.

We were amazed. Davide had somehow pushed his adult scooter over the Rockies, through the Prairies and was now in Ontario. He acknowledged that this was eccentric and said there were two other friends from Europe behind him who weren't enjoying this sadomasochistic form of fun as much as he. Davide suggested we give his Traczer a go but we couldn't even keep the thing upright, let alone dare take one foot off the ground. It was amazing to think that Davide had ridden this death trap up and down mountains. Shortly after we waved goodbye, we saw his fellow scooterists. They looked exhausted, and envious of our saddles.

Davide's interesting choice of TransCanada touring vehicle

Canada is famous for some shy but potentially dangerous animals. Outside of towns and cities you're never too far from a sign warning you about the dangers of cougars, bears and moose. These animals are certainly capable of doing you a disservice, but in reality, are more scared of you and likely to stay clear. However, Canada also has a slew of insidious insects that aren't so coy and take great delight in taking little chunks out of you and making you itch for days.

We loved Ontario. We visited all ten Canadian provinces, and it was one of our favourites. However, Ontario's army of tiny flying demons did not love us. There are horse flies, deer flies and – king of the critters – the black fly.

Whenever we were near water, we had to make sure our skin was covered. We lathered any remaining exposed skin in DEET, but the black flies laughed at our pathetic attempts to fend them off. The most frustrating thing was that they were so small you couldn't see them coming. With their tiny frames they stealthily landed on our vulnerable skin, ignoring the maximum strength DEET, and landing tiny bites that would result in grotesque swollen lumps, inelegantly protruding from our bodies.

Cycling was fine, because we could ride faster than the flies could fly, but they would catch up with us on hills and cruelly hack away at our arms and necks as we soldiered upwards. We asked several locals what their recommendation was for avoiding being bitten and the usual response was a sarcastic 'stay indoors'.

We had mixed feelings about leaving Ontario. We loved the scenery and the people, but the flies were intense and we hoped that Manitoba might bring some relief. It was with great horror that within hours of entering our second Canadian province, we found out about the Manitoba mosquito.

Everyone has a different joke about the size and propensity of mosquitos in Manitoba, but the most unsettling jibe is that the provincial bird of Manitoba *is* the mosquito. Some parts of Manitoba have embraced the insect. We didn't go there but we heard that there is an oversized statue of a mosquito just north of Winnipeg, in a town called Komarno, which is Ukrainian for mosquito.

Our first night in Manitoba was at another community campground, operated by the Lions in Whitemouth. The Lions are a community organisation that promote 'good government and good citizenship'. They also maintain a number of well run and cheap donation campgrounds across Canada. We quickly developed an affection for the Lions and it was always tempting to end the day whenever we saw one of their campgrounds.

Our first city stay in Manitoba was at Winnipeg. As we neared the city, we were cruising along a dead straight country road when the sky turned grim and a thunderstorm was clearly brewing. We hid in a gas station for two hours, along with a motorcyclist who hadn't ridden in the rain before. The downpour clearly wasn't going to stop and we wrapped up as best we could and headed into the fray. The roads in Winnipeg were no better than the weather. They were atrocious and to this day the deepest potholes we've ever seen. Our panniers were jumping around all over the place and we constantly worried about pinch punctures.

***

Less than a month into the trip we had made it to the Prairies, a vast area of predominately agricultural land that stretches from Winnipeg to Calgary, across the three provinces of Manitoba, Saskatchewan and Alberta. Our planned route across the Prairies raised more eyebrows than any other section of the trip. The most worrying concern people raised was the wind.

Most people who undertake a TransCanada bike tour start at Vancouver and head east to exploit the strong prevailing westerly wind. We were doing the opposite. Although we knew prevailing winds were westerly, we wanted to ride the northernmost part of our route in the summer months, which dictated riding west across Canada. As we rode through the Prairies, people repeatedly told us we were 'going the wrong way'. This was a valid criticism of our route and we paid the price on several occasions, in the exposed lowlands of the Prairies. With hindsight, we would have been better off completing our loop of North America clockwise. However, hindsight is a wonderful thing and had we gone clockwise we might not have had so many encounters with affable strangers. So yes, we rode into a headwind for two months, but on reflection, I'm glad.

As well as riding into a relentless headwind, we also discovered that the famously flat Prairies aren't so flat after all. There is a notorious Prairies joke that goes 'Saskatchewan is so flat you can watch your dog run away for three days'. Except Saskatchewan isn't flat and neither are the rest of the Prairies. There is a gradual incline all the way from Winnipeg to Calgary, which keeps going to Banff. On an elevation profile it looks worse than it is, but it's definitely there and most unwelcome when combined with a stiff headwind and our heavy loads.

We were also repeatedly warned about the Trans-Canada Highway, specifically Highway 1. 'Far too dangerous for cyclists', the naysayers said. On the contrary, the Trans-Canada was one of our favourite roads, with its beautifully maintained surface and gloriously wide shoulder. In fact, the shoulder is so wide on the Highway 1 bit of the Trans-Canada that you can comfortably ride two abreast and still have a big gap between you and the seldom seen traffic.

Of all the concerns people raised about the Prairies the weather, specifically tornadoes, was the most valid. Each year the Prairies experiences about forty tornadoes and we were riding through at peak tornado season. There's something about the combination of temperature and humidity that make the Prairies especially prone to them. Worldwide, only the perceptively named Tornado Alley in the US experiences more twisters than the Canadian Prairies. And we didn't escape them.

The day we left Winnipeg, the skies started to look menacing. The sort of skies we had been warned about in the next province over – Saskatchewan – who's own licence plate strapline is *Land of Living Skies*. We knew this could mean some dramatic weather was incoming, but we mistakenly pressed on all the same.

There was construction on the Trans-Canada, and we were pushed into a tight lane with the traffic. A lady in a pick-up squeezed ahead of us and into the service lane, flagging us to stop. She explained that she worked for the Canadian Transportation Agency and that there were numerous tornado warnings on her intercom. She told us we should leave the road and seek shelter immediately. We could see for miles. There was no shelter, just fields. When we pointed this out, the lady agreed that there was nothing we could do, shrugged, jumped back into her pick-up and sped off, leaving us to fend for ourselves.

A mile later and the weather had turned from menacing to alarming; a commotion of wispy black clouds swirling in the cauldron in the sky. As we were wondering whether this was what the 'calm before the tornado' looked like, a white van crashed through the traffic cones in front of us, creating its own entrance into the service lane.

The driver, who we found out was called Morton, leapt out and shouted "You need to get out of here, there's a tornado coming, get in my van."

Morton was tall and stocky but at first glance you wouldn't put him down as strongman. Appearance-wise he was definitely more Clark Kent than Superman. However, like Clark Kent, Morton had some hidden strengths that he neatly demonstrated as he rescued us from the path of the incoming tornado.

We couldn't lift our bikes with all the kit on. Whenever anyone asked us how heavy our bikes were, we'd laugh and say 'try lifting them'. Morton was not to be beaten by the bikes, flinging open the van's side door and picking them up like they were bottles of maple syrup. Aghast, we stood there like plums, not knowing what to say or do.

"Get in the front, quickly!" barked Morton.

We leapt in and the second he drove off, weather happened. Lots of weather. Rain, thunder and lightning all descending from a kaleidoscopic sky. Morton reconfirmed that there were multiple tornado warnings and that he was taking us as far as he needed to, to make sure we had a roof over our heads that night. It wasn't long before we arrived in the small community of MacGregor and found a B&B for the night. We thanked Morton profusely and he sped off to make himself safe.

The next day we were snacking on the side of the road when a man pulled up next to us.

"Where did *you* guys sleep last night!?" he asked.

We explained what had happened, and he told us that his cousin lost her barn to the tornado and was then up all night finding her sheep. The next time we found Wi-Fi we checked the news and CBC – effectively Canada's BBC – published a story with the headline '*Manitoba walloped by yet another old-*

fashioned prairie storm, tornadoes and all. The story read:

"Environment Canada meteorologists are investigating two tornadoes that touched down in the area, including one that folded Maureena McPhail's pastoral barn in on itself and put her flock of 150 sheep at risk... McPhail said she and her husband spent the evening rounding up their 150 sheep. None of them were injured. "We're quite happy about that. They're out in the pasture right now," she said. "They're upset, to say the least, but they'll be fine in the pasture until we can rebuild."

In fact, Environment Canada recorded three tornadoes in southern Manitoba that day, the first time that had happened in over two years. So yes, the weather in the Prairies is a challenge, but freak weather events can happen anywhere. As we'd go on to discover in Oregon, Florida, Virginia and Nova Scotia.

The final, and most commonly cited issue with our Prairie route, was boredom. People couldn't understand how we'd pass the time of day with nothing to look at other than tarmac and wheat. It would be fair to say that the Prairies' scenery does not rival that of its neighbouring provinces. But it's definitely not boring. Ties and I love wildlife, and we had lots of fun spotting Prairie dogs, gophers and ferrets. We were also privileged to have an early morning sighting of an endangered Swift fox, which looks like a tiny, skittish wolf.

Aside from the ubiquitous agritourism, there are also some quirky sights in the Prairies. We had a rest day in Moose Jaw, the fourth largest city in Saskatchewan with a population of 33,890 and colloquially referred to as Little Chicago. There is a tourist attraction in the town called *Tunnels Of Moose Jaw*. Nobody knows why the extensive network of tunnels was originally constructed, but the popular account is that they were inhabited by early Chinese Canadian settlers who lived underground in the late nineteenth century. When prohibition was introduced in the 1920s, Al Capone is alleged to have operated his business north of the US border in Moose Jaw, repurposing the tunnels as liquor storage. Many claim it's a myth, and even the official literature for the Tunnels Of Moose Jaw states there's no *actual* evidence that Al Capone operated there. Al Capone himself once allegedly said "Do I do business with Canadian racketeers? I don't even know what street Canada is on."

We'll likely never know if the connection is true or not, but many a similar tourist attraction has spurious origins. And who would deny Moose Jaw such a great story!

***

Before we had even left the UK, my parents had booked a trip to visit us in Vancouver. My dad spent thirty-five years in the Royal Navy and has travelled the world. My mum on the other hand had never left Europe before, with Italy being the furthest she'd ventured. Their trip was therefore a big deal to my parents. And that put us under intense pressure.

I had to calculate when I thought we'd arrive in Vancouver, factoring in rest days and mechanicals, and bearing in mind I'd never done any cycle touring longer than a few days before. By the time we made it to Calgary we were slightly ahead of schedule and decided to round off our time in the Prairies with a few rest days.

This was a great decision because we loved Calgary. We found an excellent bike shop, inventively called BikeBike, whose friendly staff invited us along to their monthly full moon ride. The guy from the shop explained that Calgary has over 400 miles of bike paths and we had fun exploring a few of them that night and over the next few days.

We also made some upgrades to our kit. We'd discovered that the flaw of our stove was that the low heat setting wasn't very low; it was in fact ferocious. That meant that if we were boiling pasta, which we did frequently, water would boil over and spit at us. We solved this by boiling pasta in two batches, but that was annoying, inefficient and costly. We contemplated buying a new stove but then discovered that we could buy a pot that's nearly double the standard size. This worked perfectly and we still use it to this day. We also invested in a folding frying pan, a blow up solar charged LED lantern and a huge mosquito net that you can hang from a tree and cover an entire picnic table with.

After a couple of days in Calgary we were well rested, restocked and ready to take on what everyone told us would be the hardest part of the trip: The Canadian Rockies.

CHAPTER FIVE

MOUNTAINS AND MENNONITES

21st August – 3rd September 2016
Distance ridden this chapter: 597 miles / 961 kilometres
Total distance ridden: 2,877 miles / 4,630 kilometres

Ski jumping is a peripheral sport in the UK. Even less popular than cycle touring. So when Michael Edwards represented Great Britain in ski jumping at the 1988 Winter Olympics, held in Calgary, it was headline news. The Brits love an underdog and Michael was the underdog of all underdogs. He was far heavier than all the other competitors, wore bottletop glasses under his goggles – which made them steam up – and was so cash-strapped he had to live in a Finnish mental hospital whilst training. The press lapped it up and renamed him Eddie the Eagle.

Eddie did not do very well, coming last in both the seventy

and ninety metre ski jumps. We Brits loved him all the more for it. I was only eight years old when Eddie rose to stardom, but I remember watching it on the telly. The Winter Olympics wasn't exactly big in our household but after Eddie's jumps we even started watching *Ski Sunday*. For a while Eddie was a British A-lister, but we Brits are fickle and enthusiasm soon waned.

In 2016 some modern day A-listers – including Taron Egerton, Hugh Jackman and Christopher Walken – brought Eddie's story to the silver screen. And Air Transat brought it to me on the tiny screen in front of me on the flight to Toronto. What a film. I loved it. I tried persuading Ties to watch it, enthusiastically exclaiming "That Eddie the Eagle film is being shown on the flight!".

"Eddie who?" replied Ties.

Whaaaaaaaaaat! Poor Eddie. He barely saw a penny from the film and in 2018 a petition to the Queen to make him an OBE (Order of the British Empire) only garnered thirty supporters.

Eventually I persuaded Ties to watch the film and she loved it. I reminded Ties of this as we headed west out of Calgary on the Trans-Canada highway.

"Don't forget to look left at the Eddie the Eagle jumps!" I yelled over the traffic, adding "Look, there they are!".

I was ecstatic. I don't think it carried the same significance for Ties, but it wasn't long before she was oohing and ahhing at the marvel of the Rocky Mountains.

We left the Trans-Canada at Morley to join the Bow Valley Trail and rode along the river to Canmore. At Canmore we found out about the Rocky Mountain Legacy Trail, a dedicated paved bike path that flanks the Trans-Canada for about twenty miles. This sounded too good to be true.

While we had enjoyed cycling through Canada, the only paved bike paths we'd found at that point of the trip were in Toronto and Calgary. We'd largely avoided the Great Trail – the 'longest recreational, multi-use trail network in the world' – put off by the rock-strewn sections of the path we'd been on. The often narrow trail is also shared with ATVs and dirtbikes, which makes it unsuitable for cycling, especially on bikes with a wide load like ours.

The Rocky Mountain Legacy Trail, on the other hand, is a joy, and quickly became our favourite bike path in Canada. As well as being paved and breathtakingly scenic, there are huge rubber mats with weight sensors that automatically open pedestrian gates as you cycle over them. The mats are also lightly electrified to deter grizzly bears and other sizeable animals from entering the path. We followed the trail to the start of the Bow Valley Parkway, a scenic road that runs parallel to the Trans-Canada.

It was late August by this point and it was cold. So cold that when we stopped to make a coffee at Johnston Canyon, I noticed ice had formed around the base of the gas cannister and frozen it to the wooden picnic table. The cold continued the next day; it was minus-four degrees Celsius (twenty-five Fahrenheit) as we rode up the steep hill to beat the crowds to Lake Louise. We weren't expecting these temperatures but it was well worth the chilly start to have the charming glacial lake all to ourselves.

From Lake Louise there are two main routes you can take through the Canadian Rockies to Kelowna. You can straight-line it through the national parks of Yoho, Glacier and Mt. Revelstoke, or head south, wiggling your way through Kootenay National Park and various provincial parks. We headed south along the quiet provincial roads, mainly as we had heard numerous accounts of the Trans-Canada becoming a bit too similar to Highway 17 for our liking.

After our twilight visit to Lake Louise, we doubled backed on ourselves to Castle Junction where we picked up the Banff-Windermere Parkway: the road that links Banff National Park to Kootenay National Park. We stopped at a rest area just before Castle Junction to brew a coffee and admire the expansive view over the Bow River. Within seconds of taking off her helmet, Ties asked me if there was something on her head. I glanced up from my coffee-making duties and was surprised to see a small brown bird proudly standing atop Ties' noggin. The bird remained planted to Ties' head as she moved around and drank her brew, only flying off when she put her helmet on.

Ties' new friend making itself comfortable on her head

Spurred on by our avian encounter, we joined the Banff-Windermere Parkway at Castle Junction to start ascending Vermillion Pass: the entry point to British Columbia (BC) and the fourth highest pass in the province. Thankfully the climb was gradual and our first night in BC was spent at the rustic Marble Canyon Campground, where we dined on the overpriced tinned stews that we'd bought from the tiny gas station at Castle Junction. As we descended from Marble Canyon the next morning, we immediately felt very pleased by the decision we'd made to head south from Lake Louise.

Within minutes of leaving the campground the Kootenay

mountain ranges came into view and we were both agog. This was the scenery we were so looking forward to in Canada. Snow-capped jagged mountain peaks flanking every road, turquoise rivers gently flowing alongside us and the comforting aroma of pine lingering in the cold, clean air. It was like riding through a brochure for Canada's Tourism Board.

Magical Kootenay National Park

Straight ahead of us all day long was Mt Sinclair (ninth highest pass in BC), our second mountain pass in a row and cycling mission for the day. By now we were becoming accustomed to cars honking at us encouragingly as we heaved our laden steeds uphill. Mt Sinclair was a beast and we had to stop frequently to refuel and refocus before the long descent into Radium Hot Springs.

The visitor centre in Radium Hot Springs was the first time on the trip that we'd been able to use a computer and it enabled me to study elevation profiles for the final push to Vancouver. It was 25th August 2016, which meant we had just over three weeks to ride to Vancouver to meet my parents. We worked out we were now way ahead of schedule.

Ties cursed me for some of the long days we'd done in the Prairies, including a ninety-one-mile punt battling mosquitoes in Saskatchewan. I cursed myself for skipping Dinosaur Provincial Park in Alberta, but I didn't beat myself up too much. Here's the thing; I discovered it's easy to work out how long it will take to ride around a continent, but REALLY difficult to pinpoint calling points. Countless times on our trip we would need to rearrange bookings or stays with people and we were incredibly fortunate for the patience and flexibility we were shown.

So what to do with our excess days? We considered arriving in Vancouver early but decided that would be too expensive: cities are not friendly on a cycle touring budget. After reading about its laid-back vibe and glorious scenery we settled on a detour to Salt Spring Island, one of the Gulf Islands between the mainland south of Vancouver and Vancouver Island. I trundled back into the visitor centre in Radium Hot Springs to plan a route on their computer.

Radium Hot Springs was where we left the Rockies, but not the beautiful mountains of BC. The crystal-clear waters and snow-capped mountains continued, seemingly omnipresent in western Canada.

We found another wonderful Provincial Park campground at Moyie Lake and pitched up. This was our well drilled campground routine:

- Be given / choose a pitch
- Ties put the tent up and set up inside
- I unloaded the bikes, hung the mozzie net around the picnic table – there was always a picnic table – and cooked dinner
- Ate half a supermarket worth of food and showered, did our laundry in the shower at the same time
- Ties washed up while I inflated the self-inflating mattresses that never self-inflated, and checked the next days' route

- Played gin rummy (final score at the end of the trip: Chris 38,395 and Ties 36,255), ate more and occasionally had a well-earned beer
- Ties normally turned in between 8-9pm and I turned in about 10pm after I'd read a bit

We had this dialled by Moyie Lake. However, Ties was taking an eternity showering that night. Eventually she came back goggle-eyed.

"I've just met a Mennonite called Birdie" Ties blurted out.

Back in the Prairies we had seen a few ladies dressed in Victorian-style dresses and bonnets. Eventually Ties was overcome with curiosity and – in Strathmore, a small town in Alberta – had complimented a lady on her attire. The lady thanked Ties and explained it was traditional Mennonite dress. We felt ignorant but resolved to find out more. We felt even more ignorant when we discovered that Mennonites originate from the Netherlands, having been named after a Dutchman called Menno Simons in the sixteenth century. Mennonites believe in complete commitment to God, are Trinitarian (believe in the doctrine of the Trinity), staunchly pacifist and non-violent.

Birdie and her husband Eugene only lived fifty miles away in Creston, which was home to a large Mennonite community. Although it was close, they enjoyed camping at Moyie Lake because it was so beautiful that it still felt like an adventure. We couldn't understand 'holidaying' so close to home but a few years later would end up buying a campervan and doing the same ourselves in Cornwall.

The next day we were going through our usual campground morning routine:

- Woke up at 6am and cursed the rain or revered the sun
- I cooked breakfast and made coffee, outside if dry or inside our tent's 'garage' if wet
- Ties packed up the inside of the tent, deflated the self-

deflating mattress that not once self-deflated, bundled up sleeping bags etc
- Smashed coffee and ate a cauldron of porridge
- Morning ablutions
- I loaded the bikes while Ties took down the tent
- Left between 8-9am

As we were loading the bikes at Moyie Lake, a man wandered onto our pitch drinking coffee.

"Hi, I'm Eugene, you met my wife Birdie last night" he said.

We started chatting, and Eugene asked us where we were staying that night. We told him that our route was taking us via the Crowsnest Highway and over Kootenay Pass. We were therefore planning a relatively short day and ending up somewhere near Creston before attacking the pass the next day.

"Perfect" exclaimed Eugene, "You can stay with us."

Encouraged by his easy friendliness and boosted by our previous experiences with the kindness of Canadians, we gratefully accepted, and were rewarded later that evening with what proved to be one of the most gratifying and memorable experiences of the whole trip.

We found Eugene and Birdie's smallholding and set up our tent in their garden, which had stunning canyon views and pens housing goats and chickens. Over a late lunch Eugene asked if we would like to attend church that evening; there was a special monthly service and they would like to invite us as guests. We made clear our secularism to Eugene and Birdie. We also weren't exactly carrying our Sunday best on the bikes. Mennonite ladies are immaculately turned out to do their groceries so we could only imagine how well they dressed for a special monthly service. This was intimidating, especially as we are both heavily tattooed. I can easily hide my tattoos under long sleeves but Ties has tattoos on her hands and fingers so cannot conceal hers as easily.

We were assured our atheism wasn't a problem and, having

cobbled together our most respectful outfits, we bundled into Eugene's SUV and set off for church. As we entered the meetinghouse – as Mennonites call it – Eugene explained that men and women sit on different sides of the room and I was to follow him while Ties was to go with Birdie. This was a tad intimidating, but we were committed by this point so wished each other luck and took our seats.

The atmosphere in the room was friendly, but austere. I looked around and realised I fit in quite well. With my dark beard, Beatles-esque mop of cycle touring hair, long sleeve base layer and plain cotton trousers, I looked like a Mennonite, albeit a scruffy one. I felt pretty smug and relaxed into the uncomfortable wooden pew.

I poked my head up to locate Ties. She did not fit in. There didn't seem to be a dress code for the men and boys. That terrible phrase 'smart casual' would just about describe it. I had that vibe going on. The ladies and girls however were decked out in the traditional bonnet, dress and apron regalia, designed to demonstrate both modesty and severance from other worldly cultures. Mennonites believe in complete separation of religion and society, so dressing in this way affiliates them with the kingdom of God, while living amongst the kingdom of the world. Given this, it wasn't entirely clear why the dress code only seemed to apply to women.

Ties was rocking her own look, very much rooted in Earthly fashion, but desperately trying to blend in with her new friends. It wasn't working. Young girls were clambering over each other to take a peek. While this was going on, the three-hour service began. It was unlike any Christian service I've ever been to. It started with a man giving a brief sermon, followed by an awkward silence, heads on craned necks looking all over the room as if awaiting something, and then someone else finally taking the lectern. This chap had a pitch pipe that he blew on lightly before bursting into song. We all joined in. A young girl came up next and performed her own squeaky but admirable sermon. I splayed my hands to clap, but luckily Eugene stopped

Mountains and Mennonites

me just in time, explaining that Mennonites don't show appreciation in that way. This was a steep learning curve.

Aside from all the ogling at Ties, the whole service felt rather solemn. When it finished that all changed. The guy behind me launched into chatty conversation, asking questions about England, our trip and if the service had made me think about the greatness of God. That caught me off guard, but I managed a "Certainly lots to think about", and we kept chatting about more grounded matters.

Everyone wanted a natter. The atmosphere changed from stifled to sociable. When we arrived back at our hosts' abode Birdie rustled up a feast of sweet and savoury delights and we talked for hours about their faith, their business and what children looked after what animal.

Birdie made us a giant stack of pancakes the next morning and, as we demolished them, Eugene persuaded us to change our route. He said it wouldn't be much fun going over the highest pass in BC (Kootenay Pass) and that Highway 3A to Crawford was a delight. We decided to take Eugene's advice and set off feeling content with our unique experience.

Eugene was right, the 3A was stunning. A gently undulating, almost traffic free road that flanks Kootenay Lake. It was also baking hot, which was a relief after the 'shiver-me-timbers' weather we'd had in the Rockies. We were both down to sleeveless tops and we even indulged in some wild swimming.

Twin Bays Beach might be the best beach I've ever been to. Powdery white sand, crystal clear water, surrounded by pine trees and snow-capped mountains. And not a soul in sight. I even contemplated skinny dipping, but my prudish British upbringing gained the upper hand and I instead contented myself with self-admonishment for such filthy thoughts. I did change without a towel though.

A few miles up from our swim, we passed the intriguing Glass House. The story behind this tourist attraction is a real head scratcher. The owner, Mr Brown, worked as a funeral director for thirty-five years, before retiring and then travelling Canada collecting 500,000 square embalming bottles, weighing a total of 250 tonnes. With no other application for them, he commissioned a two-storey house resembling a small castle. Bonkers.

We spent the night in Crawford Bay and then caught the Kootenay Bay Ferry the next morning. If Twin Bays Beach *might* be the finest beach I've ever been to, the Kootenay Bay Ferry is *definitely* the best ferry I've ever caught. A stunning trip across a lake as still as a pond. And it's free! The ferry deposited us in Balfour and we cruised into Nelson.

***

In the UK, a city is [usually] a city if it has a cathedral. Generally, this means that cities in the UK have lots of people. There are some anomalies, with St Davids in Wales being the extreme example with a population of only 1,600 people. This makes it one of only three UK cities with a population under 10,000 and one of the ten smallest cities in the world. By contrast, in Canada everywhere seems to be a city.

I imagine this is partly due to the immense size of Canada (the second biggest country on Earth by area) and the need to provide services for dispersed citizens. Cities inevitably have more amenities, but in Canada even the smallest city will have colossal grocery stores, which any European would consider to be a hypermarket; along with financial districts, museums and a campground. This makes the country excellent for cycle touring, as you know you'll almost always see a city every day, but that city is also likely to have little traffic, and to be both orderly and pollution free.

Nelson is one such city. With a population of 10,664, it's about the same size as Shepton Mallet but feels much much bigger. We loved it. The municipal campground is centrally located and, when we were there, *The Goonies* played on a massive screen on top of a bus. The town centre is quaint and wouldn't look too out of place as the set of a John Wayne

Western. We had a relaxing rest day, which we needed because the next day we'd re-join our original route on the Crowsnest Highway. That meant hills, specifically the twenty-four-mile long climb of Bonanza Pass (seventh highest pass in BC).

I desperately wanted us to make it to the top of Bonanza Pass the day we left Nelson. I knew if we did, we'd have an easy eighteen-mile descent and a proper campground at Gladstone Provincial Park to recover at. The weather had other ideas though. It rained so hard that at one point we even sheltered in the sulphurous atmosphere of a Canadian long drop. Ten miles from the summit of Bonanza Pass we arrived at Nancy Greene Provincial Park.

We propped our bikes against the shelter in the car park and ran inside to hide from the rain. After two hours I conceded we wouldn't make it over the pass that day and we decided to spend the night there, in the comfort of the shelter. We rolled out our air mattresses and sleeping bags, and bedded down. I had a very restless night. I knew we were being a bit cheeky sleeping in the shelter and that played on my mind. Eventually I nodded off but was woken up several times by a persistent mouse, who clearly thought my head made a splendiferous bed.

Ten miles into the next day we summitted Bonanza Pass and then freewheeled to Christina Lake, where the scenery completely changed. We were still following the Kootenay River but the dense green forests that we'd grown accustomed to since entering Kootenay National Park had been replaced with open dusty brown grasslands. The Crowsnest Highway that we were riding on was now hugging the US border and there were numerous signs and derelict mines harking back to the area's 'gold rush' heritage. We ended the day at Grand Forks, where we found another fabulous municipal campground. That was also the first time we stayed at a campground with fellow cycle tourers. Initially we didn't see them, we heard them. We had finished that day relatively early and were unwinding in the tent when we started to hear music and singing. Ties – always the more curious of the two of us, and excited to hear live music – ventured outside to see what was happening.

Rachel and Simon were cycle touring with a violin and a guitar, and had started an impromptu duet after their days' ride. We struck up a conversation and discovered they were also cycling 'the wrong way' across Canada, having started in Nova Scotia, but were taking a rougher route avoiding roads wherever possible.

The next day we conquered another mountain (Eholt Pass) and trickled into Rock Creek, where we hoped to end the day. The campground wasn't open though, and so we were forced to plod on. Before long we once again bumped into Rachel and Simon stocking up on locally roasted coffee. They told us they were stopping soon for the day, at Kettle River Recreation Area, and suggested we share a pitch.

Rachel and Simon cooked us a lovely dinner that evening. We reciprocated with porridge the next morning, which didn't seem very equitable but Rachel and Simon seemed okay with the rough end of the bargain. Our new friends were keen on taking the off-road Kettle Valley Rail Trail north, so we left them to it and said our goodbyes.

***

By this point in our trip, we'd been on the road for almost two months and were becoming wise to some quirky Canadian traits. One of these was a propensity to underestimate distance, which had led to a few disgruntled detours along the way. Roughly speaking, an hour in a car is a day on a fully loaded touring bicycle. We'd learnt the hard way to ignore advice like 'It's just round the corner'. The day we left Kettle River Recreation Area, we were caught out something rotten. Well, I was.

At Beaverdell, a friendly soul gave us an excitable recommendation for a campground at Arlington Lakes Recreation Site.

"Just off the main road" he pleaded.

Yeah right, I thought. Maybe he read my mind because he showed me a map. It did look close to the road. It wasn't.

My relief at seeing the campground sign was soon wiped away after the drudge of trying to ride up thirteen percent rock strewn fire roads. Behind me, Ties was having a moment. She didn't want my help and yelled at me to "GO AHEAD", so she could ascend at her own pace.

After a mile of solo riding, I resigned to a tongue-lashing and decided to wait for her to catch up. Ten minutes later, a typically enormous pick-up pulling a trailer wound its bumpy way towards me. I smiled at the driver who shrugged the internationally recognised sign for 'sorry' at me and slowly glided by. And then I saw why he was so apologetic; Ties and her bike were in the trailer. She was stood up waving at me like a proud Mardi Gras float queen.

Hands on hips, she smirked, cupped her hands to her mouth, and shouted "No room in here for you."

Fair enough, I thought. Fair enough.

Arlington Lake: worth the climb

CHAPTER SIX

DOODLEPUFFY

4th September – 2nd October 2016
Distance ridden this chapter: 501 miles / 806 kilometres
Total distance ridden: 3,378 miles / 5,436 kilometres

In the early 1900's the Canadian Pacific Railway (CPR) were having a hard time moving precious minerals and lumber from the Kootenays to the ports at Vancouver. Eventually, someone had the bright idea to route the Kettle Valley Railway through Myra Canyon, which stands at an elevation of over 4,000 feet. To cross the vast chasm, the CPR built nineteen trestle bridges and blasted two tunnels through the canyon.

Trains chugged over the trestles until 1972, before being abandoned. The Myra Canyon Trestle Restoration Society (MCTRS) formed to restore the trestles and by 1995 had converted them to walkways. In 2003 a lightning strike started a wild fire that burned for almost a month, destroying 26,000 hectares of forest and twelve of the trestles. Not to be deterred, the MCTRS began the painstaking restoration process and by 2008 had lovingly restored all the trestles destroyed by the fire. A year later *National Geographic* described 'Biking Through British Columbia's Myra Canyon' as one of the '25 Best New Trips in the World'. I humbly agree, it's spectacular.

We rode over the Myra Canyon Trestles on my thirty-seventh birthday, having been told about them at the Arlington Lakes Recreation Site. The restoration job is incredible and the trail is kept in fantastic condition. The trestles themselves are somewhat daunting, with gaps between the wooden slats, and relatively low side rails, giving you an uncensored view of the deep canyon below. If you can bear to peer over the edge of the trestles, the view over the verdant forest is unforgettable.

The trail eventually spits you out onto a near twenty percent gravel descent that snakes its way through vineyards and fruit orchards, down into Kelowna. I loved it. Ties handled the road gallantly but gripped the brakes so hard that I had to replace her disc pads by the time we reached the bottom.

***

Kelowna is the gateway to the Okanagan Valley, an area famous for being an immense sun trap and, as a result, growing a huge variety of fruit. We had a fantastic rest day in Kelowna, celebrating my birthday by sampling beers at the local microbreweries, and eating as much food as possible.

Our first night in the Okanagan Valley was spent in Penticton, where we had arranged to stay with Jamie and Patrick (friends of Kasia, who we stayed with on Manitoulin Island). Jamie is a biologist who told us all about the fascinating flora and fauna of the bizarre Okanagan Desert, while Patrick showed us his superb artwork inspired by his appreciation of Canada's rousing natural wonders. We followed our hosts' recommendations and cycled the east side of Skaha Lake which has a tarmacked road, right on the water's edge. From there we had a steep climb up to the Okanagan Desert, where wildlife warning signs changed from grizzly bears and moose to rattlesnakes and boa constrictors. Fortunately, we didn't see any.

The Okanagan Valley's contrasting landscapes of abundant orchards and arid desert seemed unfeasible. We never expected to ride around both glistening lakes and stark desert scenery in the same day. In fact, we hadn't anticipated to see desert scenery until the Southern US states. We emerged from the desert into the small village of Keremeos, which is crammed full of fruit stands that interestingly specialised in the pleasing combination of giant peaches and homemade samosas.

From Keremeos we headed to the old gold mining town of Hedley, where we came off the Crowsnest Highway to follow the Similkameen River to Princeton. The small town's name is a corruption of Prince Town and is named in honour of King Edward VII, who visited the area in 1860 whilst still a prince.

In Princeton we took a day off to prepare for our toughest day of climbing yet; forty miles of uphill to the summit of Allison Pass in the Cascade Mountains. Allison Pass wasn't particularly steep but sometimes a long relentless climb can be worse than a short, steep hill, sapping your energy all day long. The next day, we were rewarded with a forty-mile descent through EC Manning Provincial Park and down into Hope. The road to Hope winds through forests and along rivers, with a constant background of mountains making you feel somewhat insignificant.

In the 1982 film *First Blood,* when John J. Rambo wanders into Hope, Washington, he's actually in Hope, BC. Hope was chosen as the location for the classic film because of its mountain setting – perfect for the manhunt plot – and the attractive subsidies offered to the filmmakers by the BC government. The scramble to turn Hope into a US town was frantic, with billboards, mailboxes and newspaper dispensers all having to be temporarily Americanised. Nowadays the town offers self-guided tours of the filming locations – which we took advantage of – and the well-known police station from the movie is now the Canyon Golden Age Society's clubhouse.

From Hope, we followed the Fraser River through many of the other *First Blood* film settings. Coming out of Agassi, we hit the steepest climb of the trip so far; Mount Woodside Hill. Fortunately, we were braced for a tough ascent as we were prewarned by one of our Instagram followers, Ken, that it was a 'brutal climb'. Although Ken stressed how steep the hill was, it was still a shock to the system after the surprisingly relaxed gradients of the Rockies and the Kootenays.

At the Golden Ears Bridge we left the Fraser River and zigzagged through the suburbs of Surrey to Tsawwassen, where we took the ferry to Salt Spring Island. We intentionally chose the slow, scenic ferry, having been told it takes you right up to each of the Gulf Islands. This gave us a direct view of the remote rocky islands, which we didn't fancy taking our bikes onto – wary of the rough terrain – but wanted to see nonetheless. The ferry dropped us at Long Harbour and we rode to the main town of Ganges, which is a cool, laid-back place with local shops, cafes and a small but fancy marina. Like much of Salt Spring Island, our campground had a bohemian feel, with all the pitches hidden amongst the trees and an outdoor hot shower plumbed through a tree trunk.

Ties had romantic ideas of enjoying the laid-back vibes of Salt Spring Island. I had other ideas, suggesting it would be fun to explore the island by bike. I mapped a route to explore the beaches and Provincial Parks. I was expecting the stunning forest and coastal scenery, but wasn't expecting the numerous fifteen percent hills. Mount Woodside hill only held the title of Steepest Hill on our trip for a day. Relatively speaking, our ride around Salt Spring Island ended up being one of the hilliest rides of our whole trip, but Ties conquered them all including one almighty nineteen percent number.

On the ferry back to Tsawwassen we had our first whale sightings of the trip, as the captain excitedly announced a pod of orcas were swimming alongside us. I'd seen dolphins plenty of times before but it was the first time I'd ever seen whales. It was remarkable to see the huge black and white sea creatures launching themselves out of the water. Back on the mainland we refuelled and spent the night camping in a farmer's field full of horses; a unique Airbnb find. The next day we edged our way through the suburbs, back to the Fraser River. By this point we had cycled over 3,300 miles in seventy-five days. Our bikes had behaved impeccably. I'd only had one puncture and replaced one tyre. Ties had only had two punctures, replaced one chain and one set of disc pads. We were feeling invincible and looking forward to rolling into Vancouver and having two weeks off the bike with my parents. My bike had other ideas. As we cycled through Burnaby, five miles from Vancouver, my gears stopped working completely. Try as I might they would not shift and we had no choice but to push.

***

My parents had arrived in Canada while we were on Salt Spring Island. Their own trip of a lifetime had started with a ride on the Rocky Mountaineer train from Banff, and now they were installed in Vancouver. Eventually we also limped into Vancouver and found the apartment my parents had booked for us all. After an abrupt exchange of stories from both parties, I explained I needed to find a bike shop pronto. I wheeled the bike to the nearest shop where Kiefer and Nick immediately directed me to Ed, 'Canada's Rohloff expert'. I called Ed and he said I could bring the bike in the next day.

That evening, my dad, being an engineer, couldn't resist having a fiddle to work out what the problem was. There was no way I was going to let him tinker with the Rohloff geared hub. This is how Rohloff's own literature describes their geared hub:

"A planetary transmission system (or Epicyclic system as it is also known), consists normally of a centrally pivoted sun gear,

Doodlepuffy

a ring gear and several planet gears which rotate between these. This assembly concept explains the term planetary transmission, as the planet gears rotate around the sun gear as in the astronomical sense the planets rotate around our sun. The advantage of a planetary transmission is determined by load distribution over multiple planet gears."

Now my dad is handy. Very handy. He should be, he used to fix warships for a living. But I wasn't ready to let him loose on my 'planetary transmission system' with nothing more than his trusty pocket penknife. He understood but was intrigued by a silver box on the outside of my hub.

"What's that?" he asked.

By the time I'd explained that it was the shift box for the aftermarket gear levers, my dad had taken it off the bike and taken it apart. Sure enough, something was rattling inside. When my dad opened it up, two small bits of metal came out that were obviously supposed to be one.

"That's your problem" announced my dad.

The next day, Ed confirmed that annoyingly my dad was right, but all was not lost as he had contacts and would have a replacement part in time for our departure to Vancouver Island. Ed was filling us with confidence and we resolved to enjoy our time in Vancouver. Two days later, my bike was fixed.

While we were dealing with my bike shenanigans, we were also messaging one of our Instagram followers – Doodlepuffy – who had offered to treat us to an afternoon at the Brassneck Brewery tasting room in Vancouver. Free beer and a friendly face sounded awesome, but I'm inherently sceptical. Ties was intrigued. But, this was an adventure and we'd already spent the evening sleeping in strangers' beds (spare beds, obviously) so what's the worst that could happen in a public bar? Doodlepuffy had told us they were named Devon, which left us none the wiser as to whether Doodlepuffy was male or female.

When Devon arrived at the bar, we were relieved to find she was super friendly, very knowledgeable about cycle touring and knew her Mosaic hops from her Cascades. We had a wonderful afternoon guzzling delicious flights of beer, making up for an almost dry two months in the Prairies. Our meeting with Doodlepuffy / Devon also helped change my perspective and made me more open to approaches from strangers; something naturally alien to my British disposition. And something that would prove highly rewarding in a few weeks' time when we crossed into the US.

***

We had a fun time in Vancouver with my parents. We walked a loop of Stanley Park where Ties was chased by a troop of racoons, after knowingly enticing them over. We caught a ferry to Granville Island, where we scoured the Willy Wonka-esque cake stands of the market, and ate the biggest muffin I've ever seen in my life. This was all part of my parents' mission to fatten us up, having been shocked by my particularly skeletal figure. Importantly, I also had my bike back.

After three days in Vancouver, we arranged to meet my parents at the Tsawwassen ferry to head to Vancouver Island. My parents had put together an itinerary, booked all the accommodation and hired a car that would comfortably swallow our bikes. However, after a few miles of riding my gears stopped working again. I called Ed who tried talking me through the process of fixing the gears over the phone, but to no avail. Eventually Ed told us to stay put and he came and collected us in his pick-up.

Ed explained that he'd need my bike for a while to find a solution and he kindly let us leave both bikes and all our panniers in his shop while we were on Vancouver Island. I called my parents to say it was too late to make it down to the ferry and we found a cheap room in someone's house to stay for the night. The following day we bussed our way down to Tsawwassen and eventually made it to Sidney on Vancouver Island to re-join my parents for a superb holiday within a holiday.

Accosted by raccoons in Stanley Park

We had an amazing ten days. After two months of constantly planning, adjusting plans on the hop and worrying about where we were going to camp next, it was a relief to let someone else take the reins for a bit, and we relaxed into being pampered and planned for. We saw orcas and humpbacks on a whale watching trip in the Juan de Fuca Strait. We saw Prince William in Victoria (who happened to be there the same day as us), visited countless beaches and walked pristine coastal trails. We also watched the otherworldly spectacle of salmon leap at Stamp River Provincial Park. We'd seen this many times on the telly, but in real life it was mesmerising. We stayed there for hours, watching the salmon achieve the seemingly impossible by leaping into and over the weir. When we eventually prised ourselves away from the jumping fish, we had the bonus of our first bear sighting.

Throughout our trip we'd been told that if you see a black bear you should stand tall and make as much noise as possible. But if the bear is brown, hope it doesn't see you – and if it does, play dead. This bear was delicately balancing on a rock, choosing which salmon to turn into its' supper. It was a safe distance away but close enough for me to see it was definitely black. Remembering the guidance, I let out a pathetic roar and flapped my arms around manically. The bear nonchalantly turned around – looking a bit annoyed that it didn't have the opportunity to catch its supper – and trundled off. I turned round and everyone was laughing at me, having found my bear-scaring roar pretty feeble. I obviously maintain that I saved everyone's lives and was comforted to know that bears scare easier than humans.

After ten days of sightseeing, indulgent food, much-needed libations and fantastic company we parted ways from my parents back at Tsawwassen and bussed back to Vancouver to collect our bikes. Whilst we had been holidaying, Ed had been busy. The manufacturers of my faulty shift box had confirmed there was a problem with the design. Ed swapped the part for a standard Rohloff click box and an adapter that allowed a standard twisting Rohloff shifter to go on the end of my drop handlebars. It looked weird but it worked, and is still going strong to this day.

Reunited with our bikes and belongings, we rode down to Douglas and crossed the border into the US at the Peace Arch, named after the huge monument that commemorates the longstanding peaceful relationship between Canada and the US. The Pacific Coast is a popular cycle touring route so the Border Patrol Agents didn't seem too surprised to see two foreign cyclists on bikes with stuff hanging off every conceivable perch. The whole border crossing process was far smoother than crossing at Michigan. Rather than scare us about visa extension prospects, the Border Patrol Agents wished us luck with our trip and apologetically explained that we weren't allowed to carry fruit across the border. However, they did give us the option of eating it there and then. We scoffed as many bananas and oranges as we could and left the remainder to undoubtedly become their daily fruit salad.

We didn't have far to go to find Birch Bay State Park where we spent the night. We weren't the first cycle tourers to arrive that day and we weren't the last. This would become a common feature of Pacific Coast campgrounds and a significant difference to the first leg of our trip. Having only met one cycling touring couple at a campground, between Toronto and Vancouver, we'd meet fellow cyclists almost every night from Vancouver to San Diego. Another daily occurrence for the next 1,100 miles of our trip was rain. It started on our first night in Washington and barely stopped until the Golden Gate Bridge. The Pacific Coast had inspired our whole route but the stretch to San Francisco would become the most challenging month of our trip.

CHAPTER SEVEN

SUPER TYPHOON SONGDA

3rd – 24th October 2016
Distance ridden this chapter: 748 miles / 1,204 kilometres
Total distance ridden: 4,126 miles / 6,640 kilometres

No one really talked about Trump in Canada. However, in our brief stint through the Upper Midwest US, people were candid about politics and we didn't meet a single Clinton supporter. Our first stint in the USA was in July 2016, as Trump and Clinton were formally confirmed as Republican and Democratic candidates. There seemed to be a popular consensus that Clinton was corrupt and that electing another member of the political elite would exacerbate the problem with politicians. Trump was seen as the man to change that. To *Make America Great Again*, if you will.

As soon as we crossed back into the US at the Peace Arch it was clear that Trump's bid to replace Obama wasn't just marketing for the next series of *The Apprentice*. There were enormous Trump / Pence signs everywhere. Trump's team were clearly from the same school of marketing as the Vote Leave Brexit bunch. We knew how that turned out and, like the weather, it looked ominous.

It was a sobering start to the Pacific Coast, made worse by our shabby physical condition after two weeks of being spoilt rotten by my parents. Their mission to fatten me up had succeeded and we'd grown a bit too accustomed to the luxury of beds you didn't have to inflate with your own puff. Ties' immune system was fragile and she picked up a stomach bug, the consequences of which aren't conducive to cycle touring.

Fortuitously, a few days into the Pacific Coast leg of our trip, we'd arranged to stay with friends. Jon and Jess live in Renton, a suburb of Seattle and resting place of legendary axeman Jimi Hendrix. Our visit couldn't have been better timed and we ended up staying for three days, most of which Ties spent in bed. While Ties rested, Jon took me on a tour of filming locations from the cult TV series *Twin Peaks*. The highlight was the mighty Snoqualmie Falls, which appears in the TV series' opening credits. The waterfall is overlooked by the Salish Lodge & Spa; the exterior of *Twin Peaks'* Great Northern Hotel, home base for Kyle MacLachlan's character, FBI agent Dale Cooper. The next day, Ties wrapped up and we walked to the Jimi Hendrix Memorial, followed by an afternoon watching the talented fishmongers slinging slippery salmon at Seattle's renowned Pike Place Market, and exploring the artistically bubble-gummed walls of Post Alley. We couldn't be bothered to queue for a coffee from the first Starbucks outlet, but we did join the hordes of tourists at Beecher's to try out their unique squeaky cheese curds.

Snoqualmie Falls, the 'Great Northern Hotel' and Toby, Jon's adorable cockapoo

We also used the opportunity of staying with Jon and Jess to replace some misguided equipment choices. Novices make mistakes and we made plenty with our kit. Other than the bike, arguably the most important piece of equipment to a cycle tourer is the tent. Back in the comfort of our flat in London, our priorities for choosing a tent were waterproofness and security. We didn't want to be dripped on in the night and, knowing we'd be in lots of urban areas, we wanted to be able to put our bikes inside.

The tent we had gone for had a decent porch at the front and fitted our bikes in if laid flat on the ground. It was an admirable tent, but even at a vertically challenged five-foot-nine-inches, I had to limbo my way around the bikes to squeeze in and out and I kept snapping the poles. We were able to have the first few broken poles repaired in Calgary, but as I continued to snap the poles they were ripping through the outer tent and we realised we needed to upsize.

After a bit of Googling we bought a 'Big Agnes Wyoming Trail 2' online and had it sent to Jon and Jess. The inner tent is smaller than our original tent, but the front tent is massive. So big we could wheel two bikes in and store them stood up. Ties could even stand fully upright in the Wyoming and I could all but stand in the centre of the front tent. It folds up to about the same size as the original tent and is much easier to assemble. We also bought the groundsheet which would prove a shrewd purchase days later.

Our tent was a talking point at most campgrounds on the Pacific Coast. Most cycle tourers we met had tiny coffin-like tents and were much more relaxed about leaving their bikes outside. We became used to cycle tourers questioning why we had such a huge tent, knowing that deep down they appreciated the extra comfort that the weight penalty gives you when it's your home for a year.

Eventually we had to leave the luxury of Jon and Jess' home. Ties was still far from fully recovered and Jess insisted on driving us the short distance to the ferry for our onward journey to Bremerton on the Olympic Peninsula.

We had a torrid time on the Olympic Peninsula. The rain was relentless. On our first night we hid in the kitchen shelter of Twanoh State Park for three hours before giving up and pitching the tent in the rain. The next day we took the sopping wet tent down in torrential rain, rode in the rain and then put the wet tent up in the rain. This became our routine almost every day for a month.

When we decided to cycle via the Olympic Peninsula, we initially planned to ride by the Kurt Cobain Memorial Park in his hometown of Aberdeen. However, the weather sapped all enthusiasm for superfluous detours of any distance and we missed it by a few miles, just to end the day slightly early and crawl into the dry sanctuary of our palatial tent.

Washington State is nicknamed 'The Evergreen State' for good reason. It rains there a lot. However, even by The Evergreen State's standards October 2016 was particularly precipitous, setting a new record for the wettest October ever in its capital Olympia. Unfortunately for us, Olympia wasn't the only place setting weather records in October 2016 and we would go onto experience miserable weather throughout the month in Washington, Oregon and northern California.

***

Washington and Oregon are separated by the Columbia River and if you're coming down the coast there's one way over it; the four-mile-long Astoria-Megler Bridge. We had read horror stories from other cyclists about the narrow shoulders and logging trucks and resolved to hitch a ride over if the wind and rain persisted. However, as we approached the bridge our wretched luck turned, and we were treated to the bluest skies, the most glorious sunshine and the clearest views of the whole month.

Riding over the bridge was awesome. It's true that the incline after two miles is hairy, and the shoulder – which was full of dead seabirds – is very narrow. But cycling the bridge shouldn't deter cyclists from the surreal experience of being surrounded by so much water and being so close to the vast Columbia River.

The other great thing about taking the Astoria-Megler Bridge is it spits you out in the town of Astoria, where much of *The Goonies* (which I, as well as most British kids my age, had watched so many times on VHS that the tape had snapped) was filmed. The jail in the opening scene of the film is now the Oregon Film Museum (and ergo the de facto Goonies Museum). We popped in and discovered the famous truck rally scene was filmed on Cannon beach, which was on our planned route.

A few hours south of Astoria we made it to Haystack Rock at Cannon Beach, where we also caught our first glimpse of the Pacific Ocean. As someone who grew up by the sea I crave ocean views, and it felt amazing to both see the sea and smell the salty air once again.

As much as we wanted to loll on the expansive beach, we couldn't admire Haystack Rock for long. The 150mph Category Four *Super Typhoon Songda* was heading our way and forecast to cross our path in a few days. This was troublesome. Our new tent was good but we were pretty sure it wouldn't withstand a typhoon. Let alone one with a name.

We pushed on to Garibaldi, a town named in 1870 in honour of General Giuseppe Garibaldi who unified Italy earlier that year. Garibaldi's legacy is strong and to this day people around the world eat biscuits, grow beards and drink cocktails that all share his name. We didn't stay in Garibaldi to honour its hirsute namesake, but to move as close as possible to the room we had booked at Pacific City. Our night in Garibaldi was harrowing and the first real test of our new tent. The wind was so strong that we kept being woken up by the tent pushing against our faces and the poles creaking as they flexed in the gusts.

We only had thirty-six miles the next day to ride to the sanctuary of indoor accommodation at Pacific City, but it took us over five hours. Ties was despondent about the weather and within a few miles of our ride she had her second meltdown of the trip. We'd ridden around Miami Cove when we lost the shelter provided by the trees on the inlet. We were directly exposed to the gales and horizontal rain coming off the Pacific Ocean and were pushed into the traffic on Highway 101. It was quite dramatic.

Ties shouted "I can taste the sea", threw her bike down and stuck her thumb out to hitch a ride.

At that moment, no amount of pleading with Ties could persuade her to carry on for another thirty miles. Within minutes a pick-up pulled over and Ties rushed towards the vehicle. The driver jumped out and Ties thanked him profusely.

"I'm not stopping to help you" he replied, as he started securing a tarpaulin to his flatbed in a futile attempt to keep the contents of his pick-up dry.

Realising she wasn't going to flag someone down, Ties jumped on her bike and pedalled off without even looking in my direction. I can't remember how long we cycled in silence, but it felt like hours. Eventually we made it to Pacific City, drenched and grateful to know we'd be indoors when *Songda* struck.

As we woke up the next day, *Songda* had become a post-tropical cyclone and was in full swing outside. The US federal government's *National Weather Service* issued a record ten separate tornado warnings and the nasty weather caused widespread disruption along the Pacific Coast. One tornado went on to cause significant damage in the small coastal city of Manzanita – prompting the Mayor to declare a state of emergency – and the turbulent weather triggered the evacuation of forty people, from a campground 250 miles away on the north coast of the Olympic Peninsula. *Songda's* aftermath was forecast to linger for a bit longer and we extended our booking by another two nights.

Pacific City didn't bear the brunt of *Songda* but it also didn't emerge Scot free. The runway at the tiny Pacific City State Airport was completely submerged, as was the campground at nearby Whalen Island County Park. We ended up staying five nights – the longest we stayed in any one place on the whole trip – with our Airbnb host generously giving us the fifth night for free as she didn't think it was safe for us to leave.

Our down-time in Pacific City gave us the opportunity to fill out our US visa extension form, the I-539. I don't stress easily – it's a useful character trait that also makes me a half-decent consultant in my working life – but the US visa extension process got under my skin.

Being ruthlessly organised, I had packed a wad of evidence documents from home in a plastic wallet at the bottom of my pannier. We appended the evidence to the I-539's and sent them off, along with two $290 money orders from the next Post Office we found in Lincoln City. This was on the 18th October 2016 and our visas were due to expire on 15th January 2017: plenty of time, one might assume. We thought filling out the forms was a hassle, but a few months later we would discover that was the easy bit.

Battling Songda

Ten miles south of Pacific City we met a bunch of cycle tourers who were heading in the same direction. The group included a young guy called Jimbon and a couple of friends from Los Angeles called Tal and Goswin. Jimbon had flown over from Germany, bought a $100 bike in Walmart, and fashioned pannier racks out of wire and duct tape from Home Depot. I found this impressive and pondered if the eighteen months research I did deciding what pannier rack I needed may have been a little excessive. Tal and Goswin were on a two-week holiday, using their entire ten day annual leave allowance to ride from Portland to San Francisco. This was day one for them and it was so wet Tal was already talking about skipping sections by train.

We ended that day at South Bend State Park, one of many government run campgrounds that offer 'hiker/biker' pitches. The US has an incredible network of State Parks, especially down the Pacific Coast. Even at cycle pace you pass many State Parks a day and they're usually in prime coastal locations. They're also expensive and book up well in advance, which is unhelpful on a cycle tour. This is where 'hiker/biker' pitches come in. Instead of being neatly segregated and numbered pitches, 'hiker/biker' pitches are one communal pitch that only those arriving under their own steam can stay at. They cannot be prebooked, don't tend to have power and they're much cheaper than normal pitches. We stayed at many down the Pacific Coast and paid an average of just $15 per night.

The next day brought more rain followed by another 'hiker/biker' pitch in Florence, a town with a unique claim to fame: in 1970 a sperm whale washed up on the town's shore. Unsure how to remove the forty-five-foot carcass, local authorities decided to blow it up. A veteran with explosives expertise advised the authorities that twenty sticks of dynamite (four-kilograms) would suffice. The gung-ho authorities ignored the advice and elected to use 450 kilograms of dynamite. The resulting explosion threw blubber a distance of up to 800 feet, leading to a much bigger clean-up operation, damage to vehicles from falling whale debris (including, ironically, to the explosives veterans' brand-new convertible car) and ridicule from global media. In 2020, the fifty-year anniversary of the incident, locals voted to name a new recreation area in Florence Exploding Whale Memorial Park forever commemorating the strange event.

Our time in Florence may not have been as historic as the

events of 1970 but it was certainly a turning point in our trip. We woke up in the small hours with a strange sensation, like we were lying on a waterbed. It felt like we were floating: there was water underneath our aching bodies, but we were completely dry. I poked my head out of the tent's sleeping quarters and into the garage, where we tended to store everything that wasn't valuable. It had rained so hard that some of our belongings were floating. We'd hit rock-bottom; the nadir of our Pacific Coast palaver. We were very reluctant to leave the tent that morning but didn't exactly have much choice. Begrudgingly we took the tent down, strapped it to the outside of the bike rather than putting it in its bag, and pedalled off into another deluge.

A new low, as our belongings float in the tent

We left the campground and within a few miles bumped into

Tal and Goswin again. We exchanged tales of woe and Tal confessed he wasn't going to spend that night grinning and bearing it at another campground. Instead, he was going to bag a room at Motel 6.

"What's Motel 6?" I asked naively.

Tal, Goswin and Ties – all well versed in budget travel around the US – all turned in unison and threw me a blank, bemused stare that suggested I should know what Motel 6 is. I was intrigued and we all pedalled in convoy to the mystical land of Motel 6.

The four of us turned up at the reception of Coos Bay Motel 6 wetter and dirtier than any guests should have the right to enter such an establishment. To my utter delight, not only were they happy to give us a room, but they were fine with us taking our bikes in too. And they had a launderette. In hindsight Motel 6 is an American equivalent of Travelodge, but at the moment I had to concede that Tal was right and Motel 6 was the bomb.

We promptly entered the massive room with two giant beds and spread all our drenched, sodden belongings out. The room looked like a second-hand camping shop by the time we'd hung something off every conceivable hook, corner and edge. I then had a long, hot shower and while Ties was showering I hogged the launderette with all our stinking clothes.

While our dirty clothes banged about in the washing machine I started thinking about our trip. We were supposed to be having fun. Cycling and sleeping in the rain was not fun. Let's be honest, it was fucking shit. Motel 6 was cheap, but not *that* cheap with the poxy exchange rate. I opened the calculator on my phone and started crunching numbers. The 'hiker/biker' rates had been kind to our budget. I resolved to treat us to the glamour of more Motel 6's while the weather was this abysmal.

The Coos Bay Motel 6 had been a revelation but more motel stays would have to wait as the coast was becoming a bit remote and there simply weren't any motels when we needed them. Morale was high though. So high that as we rode through Port Orford I was able to tempt Ties to a detour up a steep hill that had the words 'OCEAN VIEW' painted on it. There was a guy sat on the strategically positioned bench and, while the view was indeed spectacular, we felt like we were gatecrashing his moment of contemplation. We tried not to disturb him but then unprompted, and without turning his head a single degree towards us, he said "Seen that?"

"Yes, amazing view" I replied.

"No, the grey whale."

We looked again and sure enough, there was a hulking leviathan rubbing itself against the rocks. The man explained that they do that to rub barnacles off their thick skin. We were mesmerised. The man walked off. We hung about in awe of the beast beneath us.

The rock the grey whale was exfoliating itself on

It sounds silly but the road along the Pacific Coast in Oregon rarely runs directly along the coast. It meanders inland, weaving around trees and hills. However, the section from Port Orford to the border with California is premium cycle touring. The road's smooth as glass, there's ample shoulder and you have a largely unobscured sea view.

There were many reasons why we were excited to cross into California, but number one was the weather. Our friends back home were following the news about California's record-breaking drought and assuring us that the rain would stop once we crossed the border. It didn't.

We had planned to end our first day in California after summitting the big climb out of Crescent City. According to Google there were loads of campgrounds in Klamath, on the edge of Del Norte Coast Redwoods State Park. The first campground in Klamath was closed due to flooding, as were the second and the third. Eventually we found a RV resort who let us stay. Their tenting area was also waterlogged but the kind owner took pity and let us stay on a RV pitch for a paltry $10. This was pure luxury as we had power right at the tent; usually we'd leave all our electronic devices charging in washrooms, hoping no one would nab them.

We tried ending the next day after thirty-six miles. The weather was appalling, we'd spent just $30 on three campgrounds over the past three days and were ready to splurge on another Motel 6. We came off Highway 101 at Patrick Point and pulled up to the first cheap looking motel we found. Wow, it was expensive! We tried the motel next door, but discovered that Patrick Point was a bit flash for us. We rolled on to Trinidad, the oldest incorporated town in California, but after trying three more places decided we must be in the posh bit of The Golden State. Reluctantly, we conceded we'd need to carry on fifteen miles to the metropolis that is Arcata.

We weren't able to explore the city, desperate to get indoors and seek respite from the rain. The Valley West Shopping Center in Arcata was exactly what we needed. A sodden cycle tourers' paradise of cheap motels, supermarkets and a launderette. This combination of businesses also appeared to make it the de facto meeting place for all the weed pickers that serve the abundant cannabis farms in Humboldt County. The car park was awash with hazy-eyed crusties who looked like they needed a shower even more than we did.

My phone told me that the Red Roof Inn was the cheapest establishment at Valley West. But surely not. In our state it looked like a five-star hotel. After changing my raincoat for a less wet jacket I sheepishly wandered in and asked if they had a room. Yes! I looked at the immaculate carpet I was stood on and thought they were never going to let us take our bikes in.

"And would it be at all possible to bring our bikes in please?" I said, in my best plummy English accent, while pointing at a sodden Ties who was sheltering under the hotel's overhang outside.

"Of course!" said the receptionist without hesitation, adding "Breakfast starts at 6am and is included."

Wow, I wanted to spend a week at the Red Roof Inn.

We did some much needed laundry and bought some microwavable meals and beers at Ray's Food Place. After only two beers – our first in over two weeks – we were both passed out by 9pm. We woke up in the small hours, fully clothed, lights on and TV wittering.

A few hours later we raided the all-you-can-eat breakfast buffet – a deadly offer to cycle touring guests – and shortly after that set off in the rain for Eureka. Living the dream.

CHAPTER EIGHT

BIG SUR AND BIG CITIES

25th October – 17th November 2016
Distance ridden this chapter: 847 / 1,363 kilometres
Total distance ridden: 4,973 miles / 8,003 kilometres

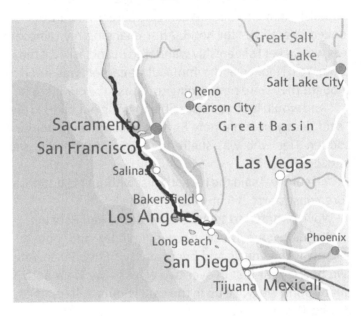

Ties has used Couchsurfing – a website where people offer their sofa as accommodation to travellers on a shoestring budget – for many years and is a big fan. She enjoyed the reciprocal arrangement, hosting dozens of guests and making friends from all over the world, in return for free accommodation on her own travels. Before our trip we had both heard of Warmshowers but I was reluctant. Warmshowers is a social media site where generous souls offer accommodation and more to passing cycle tourers. Basically, it's Couchsurfing for those on pushbikes.

The service works on the same spirit of reciprocal hosting as Couchsurfing and in 2016 was free (there is now a small fee). Some hosts offer a place to pitch your tent in their garden, while some provide a bed, food, shower, laundry and more. Back in London I couldn't imagine I'd ever be keen to reciprocate hosting after the trip and I didn't think it was fair to use Warmshowers knowing that. It's not that I'm an antisocial curmudgeon who doesn't like meeting people, I just couldn't imagine letting complete strangers stay in my house.

A few things changed my mind. Ties had been trying to persuade me to be more open-minded, pointing out that she'd had only positive experiences as a guest and host on Couchsurfing. The weather had been truly atrocious; there were only so many more mornings I was willing to wake up floating in a tent, and we couldn't afford to stay in motels regularly due to the lowly value of the Pound after Brexit. And I was encouraged by the random acts of kindness we'd experienced, from being bought beer based purely on our Instagram profile, to those who'd let two heavily tattooed oiks kip in their house.

As we rode through Humboldt Redwoods State Park, on the glorious Avenue of the Giants, I fired off my first Warmshowers request to Alison in Fort Bragg. Alison quickly agreed to host us and we would go on to stay with over sixty Warmshowers hosts over the next nine months. It changed our trip immeasurably and when we returned to the UK, we enjoyed hosting ourselves.

***

Our first Warmshowers experience with Alison and her partner Bruce was fantastic. We were offered our own bedroom, secure bike storage, laundry facilities, yummy homecooked food and interesting conversation. We were also joined by another cycle tourer from London called Joseph, who was touring the Pacific Coast solo.

Joseph's Pacific Coast tour was a walk in the park for him. Months before we met Joseph he had become the first person in the world to 'everest' on a fixie. 'Everesting' is riding up and down the same hill repeatedly until your elevation gain is the same as the height of Mount Everest. I attempted this myself in March 2021, albeit not on a fixie, and only managed 14,948ft (a 'half everest' or 'basecamp'), riding up and down Weston Hill, near Bath racecourse, twenty-seven times. Boredom got the better of me but if I'm honest I'm also not entirely sure my legs would have managed fifty-four repetitions. I can't imagine how Joseph managed it on a fixie, especially the descending. You cannot freewheel on a fixie, which means you need to spin your legs as quickly as the cranks rotate. If you forget to spin your legs at high speed, or even pause momentarily, your whole body is violently lurched forward. A thoroughly unpleasant experience.

Alison and Bruce let us stay for a second night to have a much needed rest day. We visited the local Triangle Tattoo Museum, where we spent hours exploring the colourful artefacts and chatting to the even more colourful custodians Mr. G and Madame Chinchilla. Some of the hand tools on display were truly eye-opening, as were the thought provoking exhibits ranging from traditional moko tattoos of New Zealand to the disturbing history of tattoos without consent in many countries around the world.

Bruce was a student at the world-renowned Krenov School of Fine Furniture. That evening we had the honour of a tour of the school, where some of the students were still busy painstakingly crafting exquisite furniture from rare timbers. Those that had finished for the day had started a bonfire party outside and we joined in for beers and s'mores, which involved roasting a massive marshmallow in the fire, squishing it between two biscuits with some chocolate and then trying not to burn the roof of your mouth on the piping hot confectionery.

As we left Alison's we started our first day on the famed Highway 1, better known as the Pacific Coast Highway or PCH. The PCH is hilly, mega hilly. There aren't any long climbs but there are lots of creeks and rather than bridging over them the

Big Sur and big cities

road engineers elected to snake the PCH around them. This means that every time the PCH meets a creek there is a steep descent left followed by a hairpin turn and a steep right hander skyward.

The day started innocently, riding along beachside boardwalks and enjoying the sea view. After a few miles we met a cycle tourer from New Jersey called Brent who, despite his young age, we'd go onto discover was a veritable Swiss army knife of a man. Brent was aiming for Manchester State Park and we said we'd meet him there, bumping into him repeatedly during the remainder of that day. We went on to form a good friendship with Brent and would end up leapfrogging him all the way down the PCH, staying with his parents in New Jersey and spending a night with him on a schooner in Maine.

Back on the PCH and we hit the first of many hills as we traversed the road snaking around Navarro River, Greenwood Creek, Elk Creek and Irish Creek. Exhausted, we made it to Manchester State Park which was shut due to flooding from all the rain. Brent turned up soon after and said he was pushing on to the campground at the small town of Gualala, whose name is a corruption of its original Pomoan (a First Nation language) moniker *ah kha wa la lee* meaning 'coming down water place'. Given the weather, this seemed a fitting name for the small town, but the Bokeya people (now the Manchester Band of Pomo Indians of the Manchester Rancheria) settled on *ah kha wa la lee* due to its prominent location at the mouth of the Gualala River and Pacific Ocean.

I had a look on my phone; Gualala was over twenty miles away and it would be dark in less than two hours. We had decent dynamo powered lights but had a strict no cycling in the dark rule, partly due to Ties' poor eyesight in the dark, and partly due to a fear of drunk drivers. With all the hills, we were averaging less than nine miles an hour so we weren't going to make it to Gualala in the light. We wished Brent well and hoped we found somewhere not listed on Google Maps. No such luck.

The final twenty miles of the day were tough. The PCH was

up and down the whole way, there was relentless rain and nagging uncertainty about where we'd stay or if we'd make it there in the light. As we reached Anchor Bay, only a few miles from Gualala, it became pitch black. We were in a wooded area by now and in the eerie darkness we could hear rustling all around us in the trees. As we ascended the last hill to Gualala there was a loud snap on our left and Ties was almost knocked off her bike.

"What the fuck was that?" she shouted, trembling and trying to keep her bike upright on the steep gradient.

Two deer had burst out of the woods and darted right in front of Ties, narrowly missing her front wheel as they dived back into the woods. Ties was shaken, and in all the commotion, her chain had come off. I laced her chain back on and we drudged to the top of the hill, before coasting into Gualala where thankfully we found a grocery store and a friendly B&B.

***

Not long after I met Ties in May 2015, Ross also started dating Harriet, now his wife. Ross and Harriet were keen to come and meet us somewhere on the trip and I thought it might be fairly easy to predict our arrival to San Francisco. We arranged to meet there in early November 2016. In Oregon, *Super Typhoon Songda* had put us behind schedule, and I warned Ross that we'd go as fast as possible but we might not make it in time to meet up.

Ironically, it was the rain that ended up pushing us there on time. We didn't feel inclined to spend rest days in sodden campgrounds and after Pacific City we rode ten days and over 500 miles on the trot. By the time we arrived in Gualala we were ahead of schedule and could take it easy to time our arrival in San Francisco with Ross and Harriet's visit. We decided to have a few short, quiet days enjoying the views on the PCH before the final few miles through Marin County.

As someone who caught the cycling bug through mountain biking in the early 1990s, Marin County has a hallowed status. While debates still rage today as to who invented mountain biking, not many argue that the sport originates from Marin County, specifically Mount Tamalpais. The Larkspur Canyon Gang started riding heavy balloon-tyred single speed bikes down Mount Tamalpais in the late 1960's. What started as a group of friends having fun on ungainly bikes eventually became a worldwide phenomenon and a multi-billion pound industry.

My desire to see the origins of mountain biking made choosing our route from Point Reyes Station to the Golden Gate Bridge a hard choice. We could take the iconic but hilly and shoulderless route over Mount Tamalpais. Or we could go around the mountain and visit the Marin Museum of Bicycling at Fairfax. After the torrid time we had riding to Gualala we didn't fancy unnecessary elevation and opted for the latter. This had the added benefit of a fantastic stay with hosts Margit and James, in Lagunitas.

Margit and James have been cycle touring in over thirty countries, logging over 75,000 miles. They have fountains of knowledge and awe-inspiring stories to share. Our favourite yarn was how they rode from San Rafael, California, to Buenos Aires, Argentina in 2013-2014, successfully evading the hostilities of Darién Province, a vast swathe of forest that separates Panama from Columbia. Buried within Darién Province is the Darién Gap, an inaccessible sixty-six-mile stretch that has become popular terrain for armed guerrillas and drug traffickers. Several high profile kidnappings have all but blacklisted it as a travelling route and crossing the Darién Gap has long been a cycle touring headscratcher. Most travellers catch a small plane from Panama but Margit and James are adventurous folks and elected to tackle the high seas, taking a *lancha*, a small fishing boat, from Puerto Obaldía in Panama to Capurganá in Colombia. It was highly entertaining, although very unnerving, how casually Margit and

James told this story, given how potentially dangerous the crossing obviously was.

After regaling us with cycling stories, James was itching to ride his bike and offered to ride with us to Sausalito the next day. The route took us through some lovely neighbourhoods, that we almost certainly wouldn't have found on our own, on a combination of cycle paths and quiet roads. Sadly, the museum at Fairfax was closed but the weather was glorious and nothing could dampen our spirits after it was finally sunny again. Ties had long been saying that the weather would change at the Golden Gate Bridge and thankfully she appeared to be right.

The Golden Gate Bridge is infamous for being shrouded in fog and it's rare to snap a clear photo of the illustrious landmark. After our month of rain, we were delighted to be rewarded with a completely clear sky.

We rode over the bridge, slaloming the throngs of tourists admiring the views of San Francisco Bay and Alcatraz. Once on dry land we stopped to ask a couple to take the obligatory photo of us: the very same photograph that now adorns the cover of this book. I recognised a familiar lilt to their accent. Sure enough, they were from Hessle, the neighbourhood in Hull where my mum grew up. Small world.

We braced ourselves for San Francisco's legendary steep streets and rode to Golden Gate Park to meet Ross and Harriet. We had a fantastic time in the city, hanging out with our friends, exploring the colourful neighbourhoods and attending the Día de Muertos (Day of the Dead) celebrations. We also coincidentally bumped into Brent in a supermarket and made vague plans to meet up at Half Moon Bay a few days later.

We stayed in a beautiful Victorian house in San Francisco owned by a wonderful host called Cyndi. Our room felt like an entire apartment, with the added bonus of a bike shop's supply of tools and lubes. Cyndi also had a creative recycling system, whereby guests could leave unwanted items of clothing, bike parts or camping equipment in a box and in return help themselves to items left by previous guests. Such a thoughtful

Big Sur and big cities

idea.

We spent three nights in San Francisco but it felt much longer and it was hard to prise ourselves away from a city we both admire fondly. It had been fantastic to hang out with friends and Ross and I reminisced about the evening in The Rake three years earlier, when he had planted the travelling seed in my head.

***

Sure enough, the night we left San Francisco, we met Brent at Half Moon Bay. We arrived fairly early and were chilling out in the tent when we heard rustling. I poked my head into the tent's garage and there was a gang of young racoons raiding our panniers. As I stepped into the garage all but one of the racoons scurried off. I made several unsavoury hand gestures and shouted something unpleasant at the little bandit, but it was unfazed. I lurched at the racoon, and the cheeky imp just scowled back at me.

Eventually the racoon found what it was looking for and sauntered off. At 4am the racoons returned for round two but our food panniers were safely in the bear lockers by then. We resolved to pay closer attention to our food panniers and stash them in the bear lockers, that all State Park campgrounds had, as soon as we arrived.

Racoons were not the only mammals trying to pilfer our belongings. When you stay at campgrounds with bear lockers there is normally a sign – and if not it's an unwritten rule – saying not to lock the lockers. This is to enable as many people as possible to cram their food in there and to prevent issues with people leaving at different times. Sadly, this also makes it easy for people to steal your stuff. We stayed at a campground in Monterey where we were constantly woken up by the sound of people rifling through the bear lockers in the middle of the night. Our bag was black and had been pushed right to the back so had gone unnoticed but sadly several other cyclists had their panniers and kitchen equipment stolen.

***

Before we started our trip there were a few segments I was especially looking forward to riding, namely the Canadian Rocky Mountains, Florida Key's Seven Mile Bridge, the Cabot Trail in Nova Scotia and the mountainous stretch of California's coast known as Big Sur. On 22nd July 2016 – whilst we were cycling through Wisconsin – some idiot wandered into Garrapata State Park and started a fire that almost took Big Sur off every travellers' wish list. The culprit has never been caught but the fire burned for three months and took 5,000 firefighters and $260 million to contain, at the time making it the most expensive wildfire to fight in US history.

The dodgy campground we stayed at in Monterey was full of cycle tourers heading south and – other than the theft – there was but one topic of conversation; to ride Big Sur along the coast or detour inland via Highway 101. All the rumours suggested the coast was covered in ash and was impassable. Additionally, a number of cycle tourers were adamant that all the campgrounds were closed and the one shop in Big Sur wouldn't be open.

The day we left Monterey felt like any other, but it wasn't. It was 8th November 2016, the day that sixty-three million Americans voted Donald Trump as their new leader. We couldn't vote, obviously, but if we could it would have been a much easier decision than which route to take. We decided to head to town and do some laundry, giving us time to try to find some more information. We couldn't find any. Ultimately, we decided that we weren't going to spurn the opportunity to see Big Sur on hearsay and the worst that could happen would be we'd hitch a ride back to Monterey if everywhere was shut. So we headed for Big Sur along the coast.

After only twenty miles we made it to the spectacular concrete megastructure that is Bixby Creek Bridge. Built in

1932, the bridge spans a canyon and was designed to enable residents of Big Sur to access Monterey when the inland route was impassable. In 1932 it was the highest single-span arch bridge in the world and as we cycled over it, in 2016, it still felt very high to us.

Squint and you can see Ties waving on the epic Bixby Creek Bridge

Ten miles later we arrived at Andrew Molera State Park, which mercifully had a large 'Open' sign on the gate. It was looking like we'd made the right decision. As an added bonus, everyone else seemed to have been put off by the rumours about Big Sur closures and there was hardly anyone else at the campground. Andrew Molera State Park is a relatively primitive State Park, owing to the stipulations made by Frances Molera who sold the land to the authorities in 1965. Frances put clauses in the deeds requiring the land to remain relatively undeveloped and the State Park to be named after Andrew, her late brother. Andrew was responsible for introducing artichokes to Americans in the early twentieth century. He started growing the bulbous vegetable on his family's land in the Salinas Valley, near from the State Park that now bears his name. Today the Salinas Valley remains hugely important to the artichoke, with ninety percent of the country's yield coming from its fertile soil.

As the sun descended, and we cooked our regular pasta dinner, another cyclist rolled into Andrew Molera State Park. Anthony had ridden the Great Divide – an off-road cycling route from Canada to New Mexico – and was making his way back north on tarmac. We starting chatting and soon realised that we had common ground. Anthony was an English doctor who had worked at Derriford hospital in Plymouth with someone I knew. Small world indeed.

Reeling from the news that Trump had won the presidential election, the next morning we set off to stock up at the Big Sur General Store, which we knew would be the last shop we'd see for two days. By now we'd realised that the best lunch carb was wraps. Loaves of bread were useless, as they became squashed and hopelessly misshapen in our panniers. Wraps were perfect as you could roll them up and stuff them in a gap with little consequence. Unfortunately the shop was out of wraps. However, the owner said he might have a solution and five minutes later re-emerged with a wad of dustbin lid-sized wraps.

"These are from our restaurant" he explained "You can buy these if you want?"

And so, after demolishing a few for breakfast, we set off to ride up Big Sur with twenty twelve-inch wraps secured to our panniers with bungee cords.

Big Sur was everything we hoped for, and from the PCH there was very little sign of the devastating three-month long fire that had only been contained two weeks earlier. The hills were tough but the unobscured views of the rugged coastline were the most dramatic I've seen outside of Cornwall. We were also fortunate to bump into a local couple having a picnic on the side of the PCH enjoying the view. The couple lived in the mountains and were taking their immaculate 1959 Austin Healey Sprite out for its annual spin. The car was canary yellow and their two chihuahuas were wearing matching coats, sheltering from the sun under umbrellas.

As we left Big Sur, we had our eyes peeled for wildlife. We were told there was a free viewing platform at the Piedras Blancas Rookery where we could expect to see elephant seals. As we approached, we heard a ruckus from our right and in the river that leads to Arroyo Del Corral Beach there were hundreds of the creatures. The enormous males were fighting, taking lumps out of each other while the females looked on wistfully from the wings. We imagined what the busy viewing platform would be like and felt privileged to be travelling by bike, which affords unique opportunities like this private viewing.

Some of the battles going on at Arroyo Del Corral Beach

The elephant seals were amazing, and definitely a first for me, but it's the sort of wildlife I'd hoped to see on the Pacific

Coast. What I didn't expect to see in California were zebras. In the roaring twenties William Randolph Hearst, a publishing tycoon, built his dream house – a castle – and created the world's biggest private zoo on land he inherited from his dad, a mining zillionaire. Hearst had all sorts of weird and wonderful creatures brought over, including kangaroos, giraffes, camels and zebras.

The Hearst Garden of Comparative Zoology, as the zoo was elaborately titled, closed in 1937. However, the zebras remained and are now left to roam 'free' and find their own fodder in 83,000 acres of land beneath the castle. We spotted a dazzle of zebras gayly bouncing around as we rode along the PCH. That was enough excitement for us and we ended the day in San Simeon, relieved to find a cheap motel after a few days without a wash.

South of San Simeon we took several detours from the PCH to enjoy the quiet farm roads. We rode up the challenging Harris Grade Road through the Purisima Hills to avoid the congested highways around the sprawling Vandenberg Air Force Base. This gave us a breath-taking view of the military facility which has the lofty function of operational command centre for the US Space Force. As we descended into the town of Lompoc, we stopped at the abandoned Valley Drive-In Theatre to take pictures and refuel. A few years later, Ties spotted the wall in the awkward 2019 comedy film *Paddleton* and we reminisced about the lunch we had there in warmer climes.

The famously sunny Southern California weather accompanied us as we meandered along the dedicated bike paths in Santa Barbara, where we had a relaxing rest day with our friendly host Duncan. In the fourteen months prior to our visit, Duncan had hosted over 200 cycle tourers! With such a prolific hosting history, it was unsurprising that Duncan was completely unfazed by our visit but incredible how eager he was to hear our stories and give us advice.

The Valley Drive-In Theatre from 'Paddleton'

It's 100 miles from Santa Barbara to Los Angeles (LA) and you can ride almost all of it on dedicated bike paths. The route takes you past numerous beaches and hugs the coast the whole way. LA however, was a conundrum. We had friends in the city that we wanted to see, but we were also reluctant to cycle around the notoriously congested city. We booked the cheapest accommodation we could find in Venice so that we didn't have to cycle too far from the coastal bike paths. As it turned out we needn't have worried, LA was a revelation.

Despite previous reservations, we spent a rest day on our bikes riding around Beverly Hills and West Hollywood. We soon discovered that LA has a system; roughly speaking every other road in the city grid is a 'cycling road'. If you happen to be on the 'wrong road' woe betide you, but follow the signs and it's an easy city to cycle around. The next day I made a few mistakes with our onward route.

We'd booked some accommodation in Long Beach, just south of LA, and set off via the idyllic bike paths that dissect LA's pristine beaches. I decided it would be fun to ride via Point

Vicente through the well-to-do millionaires' playground of Palos Verdes Estates. It was fun, until we reached the Port of LA. The US has vague rules around what you can and can't cycle on. For example, as we would come to discover, Interstate Highways (equivalent to UK motorways) are a no go for cyclists unless there is 'no feasible alternative'.

The Vincent Thomas Bridge from the Port of LA to Long Beach is not vague; its unmissable 'Bicycles Prohibited' sign is self-explanatory. This forced us onto the service road for Highway 101, one of the most terrifying roads I've ever ridden on. Huge port vehicles buzzed passed us at breakneck speed with inches to spare. The tarmac was riddled with deep potholes, full of puncture-inducing debris. It was a horrible but mercifully short section of road that spat us out onto the bike path alongside LA River. 'Ah, bliss', we thought. Nope! We rode a gauntlet of intimidating groups of miscreants intentionally blocking our path. At one point we were being crowded so tightly, that we had to use our bikes as battering rams to clear the path.

Our time on the dodgy river was also mercifully short and we were soon in the room we'd booked. We did some groceries and collapsed in a heap on the sofa. Minutes later we heard loads of shouting and banging.

"What now?" I said to Ties.

Ever curious, Ties jumped up to see what was going on outside. It was an anti-Trump rally. Now this was fun. Some of the outfits and placards were genius. My personal favourite was 'Republicans for Voldemort', a reference to parallels between Trump's White House and the malevolent baddie from the *Harry Potter* franchise.

Months later, in February 2017, the website Buzzfeed ran a humorous quiz called 'Who Said It: Steve Bannon Or Lord Voldemort'? Although funny at the time, Trump's policies of separating and caging migrant children, banning Muslim visitors and withdrawing from the Paris accord would prove to be very Voldemortian indeed.

CHAPTER NINE

SUPERTRAMPS

18th November – 2nd December 2016
Distance ridden this chapter: 560 / 901 kilometres
Total distance ridden: 5,533 mile / 8,905 kilometres

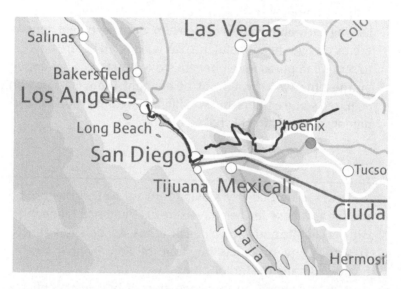

Just north of San Diego there's a massive military base called Camp Pendleton. If you're a US citizen, and you're very organised, you can apply for a pass to ride a bicycle through Camp Pendleton. Everyone else needs to revert to Interstate 5.

Like the Trans-Canada, riding Interstates in the US feels odd at first, like you're doing something wrong. But the huge North American shoulders make cycling on major roads a doddle. We ended our stint on Interstate 5 at Cardiff-by-the-Sea and quickly found San Elijo State Park, the busiest and most expensive campground we'd seen at the time. It was madness, with the usual variety of camping vehicles all crammed into tiny pitches.

Thankfully, they offered 'hiker/biker' rates but even these were double the usual cost and we had to plead to secure a spot, keeping shtum about the lavish size of our tent.

Brent was one of many cycle tourers we saw at San Elijo. He was very pleased with himself, having made sourdough bread from scratch in a stove he fashioned out of an American-sized baked bean tin. Brent explained that he'd been given the fermented culture from another cyclist and was now carrying it around on his bike like a baby, waking up to freshly baked bread every morning. We retired to bed wishing Brent well as we were leaving early and wouldn't catch him. Brent said he'd definitely see us as, ever resourceful, he was waking up before daybreak to scour campfires for embers to stoke his stove. Sure enough, by the time we crawled out of our tent Brent had not only acquired some embers, but was filling the campground with the homely aroma of freshly baked bread.

***

Ties was particularly looking forward to San Diego. She'd used it as a base for her travels in 2013 and had fallen in love with its climate and 'life's a beach' ethos. As soon as we arrived, Ties took me on a cycling tour of San Diego's beaches, boardwalks and the famous pier.

It only dawned on us as we started brewing a coffee on Ocean Beach that we had completed leg two of our trip. We'd ridden over 5,000 miles through five Canadian provinces and six states in 140 days. As we clinked our enamel camping mugs a young couple, Scott and Melissa, started asking us about our trip. After chatting for a few minutes, Scott invited us to spend Thanksgiving with his family in the hills of Ramona.

The origins of the American event are contentious and somewhat controversial. The oft taught version is that the day celebrates the arrival of the Pilgrims in Plymouth, Massachusetts in the seventeenth century. The story goes that the Pilgrims spent three days breaking bread with the

Wampanoag First Nation people as a sign of unity and gratitude.

In recent times, this interpretation has come into question and doubts have been raised as to whether the Wampanoag were invited to the party at all. Some go further and since 1970 have observed Thanksgiving as a National Day of Mourning, in protest of the US' treatment of First Nation people.

We had no plans to celebrate Thanksgiving but were always keen to meet local people and experience cultural events. We thanked Scott for his invite, making certain his mum didn't mind the intrusion of two smelly European cyclists, and arranged to meet two days later.

Scott suggested we bring a dish to Thanksgiving, so we spent the following day exploring the exciting international grocery stores and markets east of San Diego. The stores seemed to offer provisions from seemingly every country on Earth and Ties rustled up a cycle tourers' take on a traditional Moroccan couscous salad.

Scott insisted on picking us up at the bottom of the Ramona Hills, explaining that it was a steep climb to his family's house. As we set off in his pick-up, Scott explained that as Thanksgiving is an important family day there was a strict no politics rule. Scott and his dad apparently had conflicting views about Trump and he didn't want this topic to raise its ugly head.

Our relief quickly dissipated as I mistakenly chose to endorse Scott's request by replying "Of course, I can understand it must be awkward that your dad supports Trump."

To my surprise, Scott replied "Actually, I'm the Trump supporter."

Now this was awkward. Scott had gone out of his way to invite us to a special family get-together and I had put my foot in it royally. I had presumptuously stereotyped a well-travelled Californian in his early twenties as being a Democrat. Things became tense and eventually Scott explained that he was a military medic and had voted Trump for one reason only;

his God-given right to bear arms. I found this a tad blinkered, especially for someone who presumably saw first-hand the effects of guns. I tried to recover the situation by pitifully agreeing that if that was his view, Trump was indeed the right choice for him.

All four generations of Scott's family gave us a warm welcome and presented us with an immense selection of festive dishes. There were familiar dishes on offer, as well as some novelties such as the surprisingly scrumptious roasted sweet potato with marshmallows. We ate and drank with our new friends all day, ensuring we were appropriately fuelled up for the start of our Southern Tier leg the following morning.

Although there was wide acknowledgement of the no politics rule, it was clear that most of Scott's family did not share his appreciation of the President-Elect. Scott's grandad was particularly cheeky, winding him up at any given opportunity. Later in the evening, after a fair bit of alcohol had been consumed and most people had retired to bed, Scott's friend Geoff, a fellow Trump supporter arrived. The gloves were off and politics was back on the table. The chief topic of discussion seemed to be the second amendment and the importance of gun ownership.

The uncomfortable atmosphere was intensified when Geoff posed the question "but what would be the best gun to use in the event of a zombie apocalypse?"

Assuming this was a joke, I let out an involuntarily howl of laughter. This was met with icy stares and a painful silence, that was only broken when Scott actually answered the question and the two had a genuine debate about the subject while their partners sank further into the sofa. We went to bed soon after: to paraphrase the poet Lemn Sissay, we needed to get out of there like penguins in a hot tub.

***

Ties had driven through the desert before and was well

aware how long it would take to cross the dusty landscape by bike. Aside from our brief foray into the Canadian desert, I had never been anywhere like the southern states and was intrigued about what the next few months had in store. We would come to discover that the desert is not an endless dustbowl, and has an incredibly varied landscape, which you might not appreciate as much while road-tripping by car.

Day one in the Southern Tier started with a huge climb to the top of Volcan Mountain, followed by a glorious freewheel down Banner Grade Road. We were in the Anza-Borrego Desert and couldn't believe we were only fifty miles from San Diego. The desert landscape was stark and eerily quiet but much greener than I expected. By late afternoon we were pleased to secure a spot at the Tamarisk Grove campground, particularly as we hadn't seen any opportunities to acquire water.

Our route directly out of the Anza-Borrego Desert was dictated by two major factors: the Salton Sea and the 2007 Emile Hirsch film *Into the Wild*. The Salton Sea is 236 feet below sea level, making it only marginally higher than Death Valley and one of the ten lowest points in the world. It's also a lake (not a sea), in a desert, and who wouldn't want to see that!? *Into the Wild* is a biopic about the precocious Christopher McCandless, who shuns his preordained career in law to hitchhike across America. We are both fans of the film and therefore couldn't resist a detour to Slab City; one of the film's iconic settings.

As we left Tamarisk Grove in search of the Salton Sea the desert became very noisy. This was odd, as the previous day had been so quiet and at night all we'd heard were the cries of coyotes. However, on day two of the Southern Tier we were unwittingly riding through the Ocotillo Wells State Vehicular Recreation Area. In other words, this is where Californians come to play with ATVs of all shapes and sizes imaginable. It was like being on the set of *Mad Max* and while it was exciting at first, we quickly grew tired of the incessant buzzing and the noxious fumes. We found out days later that if we'd ridden one road parallel, a few miles away, we could have avoided Ocotillo Wells and seen the giant metal animal sculptures at Galleta Meadows. We resolved not to dwell on this and 'put it on the list' for future visits to California, of which we were certain there'd be many.

Thankfully, the detour to the Salton Sea turned out to be a great decision. We stayed at the white sand Mecca Beach Campground, which had spectacular views of the Santa Rosa Mountains over the glistening Salton Sea.

The Salton Sea, white sand Mecca Beach and imposing Santa Rosa Mountains

At this point in the trip we hadn't done any wild camping. From Toronto to Vancouver, it simply wasn't necessary as there was such a frequent supply of high-quality campgrounds. From Vancouver south we had started to retreat to motels due to the weather and we'd discovered the delights of Warmshowers. Ties was nervous about wild camping in the US for fear we would unintentionally pitch up on someone's land, and the potential for such matters to be settled with a revolver; we'd seen many 'We don't call 911' signs on the trip. I was less worried about guns, as whilst I was aware most people carry, my logic was potential assailants would also assume that we were carrying and therefore not give us any bother. I was more concerned about having a shower after riding fifty miles (which was our average daily punt over our year on the road) on a bike weighing fifty-five-kilograms.

We knew sooner or later that we'd need to do some wild camping and we decided we'd start in Slab City. In *Into the Wild* Christopher McCandless (or Alexander Supertramp as he likes to call himself) makes Slab City his temporary home before readying himself for the wilderness of Alaska. Nowadays, Slab City is a temporary home to snowbirds – those who flee colder climes during winter – and permanent home to free spirits who want to live off-grid. The land was once a military training base and while technically now owned by the State of California, it is considered by those that live there to be the 'Last Free Place in America.'

It wasn't at all what we expected. Some of the better known sites – Salvation Mountain and the whacky East Jesus sculpture garden – are definitely worth seeing. But considering that everyone who lives there is effectively squatting, its odd that there are KEEP OUT and PRIVATE PROPERTY signs everywhere you turn. The original ethos of the anarchic drifter community seemed to be lost on those that have cordoned off large areas of Slab City to inappropriately claim as 'theirs'. To us, Slab City resembled an unofficial, downtrodden trailer park, full of makeshift houses and rusty motorhomes. Reluctantly, we found a small unclaimed area of land and set up the tent amongst all the rubbish and burnt out vehicles.

From Slab City we filled up with groceries at Calipatria – claim to fame, tallest flagpole in the US – and then left the highway to ride through the farms of Imperial County. We were heading into the North Algodones Dunes Wilderness Area and, as the name suggests, knew there wouldn't be many amenities for the next few days.

The dunes were stunning and all we could see as far as the horizon were perpetual waves of sand. As it became dark, we started looking for somewhere to camp. It was obvious we weren't going to find a campground before sunset and eventually we found a toilet block just off the road with a conveniently flat concrete patch nearby. The only issue were all the ATVs buzzing around. They didn't seem too selective where they went, so we had a nagging worry as to whether we'd be flattened in the tent. We decided to pitch the tent as close to the toilet block as possible, hoping they would stay clear of a concrete structure.

Moments after we crawled into our sleeping bags a police car pulled up next to us, its flashing lights turning our tent into a disco. We lay there silently, praying to the Gods of Cycle Touring that they weren't going to move us on. Our fear that we'd done something wrong turned to annoyance as the lights were giving us a throbbing headache. Our unjustified annoyance then turned to mild concern when we overheard on the police radio that the officer was in the dunes looking for an escaped prisoner from Calipatria State Prison. Thankfully the escapee either didn't come our way, or wasn't interested in whatever could be inside a tent, and we lived to tell this tale.

***

There are some things that don't have much coverage in the cycle touring books I've read, in particular the tricky business of number twos. We carried toilet paper with us but by this point had always managed to find a bathroom somewhere. As we cycled through miles of enormous dunes we weren't hopeful we'd find something. It had become a running joke between Ties and I as to who was going to answer natures call outside first. On that day in the desert, as our stomachs cramped and groaned, the goading between us became incessant.

We left the dunes and entered an altogether rockier

landscape. As we admired the red rock faces of the mountains, the body clocks were ticking. Then, like a mirage, in the proverbial middle of nowhere, a portaloo appeared. We rejoiced at its unexpected cleanliness and then surmised that it must have been donated by a cycle tourer who'd previously passed through the desert.

For the third night in a row, we wild camped that night. We hadn't intended to, but it turned out that Palo Verde County Park wasn't a campground after all and we found a secluded clearing amongst the trees to pitch up for the night. It was a great spot on a small lake and there was an immaculate toilet nearby, with power to charge all our electronic devices.

By the time we reached Blythe, we'd spent almost six weeks in California, and our last night in The Golden State was a real treat. We stayed at a bait and tackle store that the owner Wayne lists on Warmshowers. After washing off three days of sweat, sand and muck Wayne ordered us pizza and invited us to hang out with him and his mates. We had lots of fun sharing unprintable stories and beer with Wayne's friends, including the affable Mayor of Blythe.

***

We spent most of our first day in Arizona on Interstate 10, peeling off momentarily to have lunch in Quartzsite, 'The Rock Capital of the World', home of the world's largest belt buckle and setting for the 2021 film *Nomadland*. In the film, Frances McDormand's character *Fern* travels to Quartzsite to visit Rubber Tramp Rendezvous (RTR), the Glastonbury of RVing.

The inaugural RTR took place in 2016, the same year we visited Quartzsite. It was clear then that the town was well set up for camping, but sadly it was too early in the day for us to pitch up. We did however discover the magic of Glacier machines, which are vending machines for tap water. For a measly Quarter (about twenty pence) we could comfortably fill all of our tiny water vessels, while the nomads in the queue looked on in bemusement, waiting to fill their giant wheeled water barrels.

That evening we wild camped again, this time on the side of Highway 60 at Desert Wells. We'd been told that wild camping was permitted in Arizona, as long as it was on public land. This didn't make the experience any less unnerving and we made a futile attempt at hiding our massive bright orange tent amongst the low desert scrub.

Fortunately, the road was very quiet and we slept well. After our usual hot-weather breakfast of wraps filled with banana, jam and peanut butter, washed down with Folgers coffee, we continued on Highway 60. We were flanked by the stunning Harcuvar Mountains Wilderness to our north and the imposing Harquahala Mountain to our south. We were also surrounded by endless cacti.

For us, one of the marvels of the desert is seeing all the weird and wonderful varieties of cacti. In Arizona you don't have to go looking for them: they're everywhere. The towering, tree-like Saguaro cacti provide endless entertainment with their contorted arms resembling whatever your imagination can conjure. We also saw lots of Prickly Pear cacti, which resemble a human hand with green fists and stumpy purple fingers.

We loved seeing the eccentric cactus varieties. Unfortunately, the desert is also full of plants that drop 'goatheads'; seed pods with horn-like protruding spikes. Goatheads don't care if you have 'puncture proof' tyres, they enjoy the challenge of penetrating your rubber hoops. Although they look quite big close-up, Goatheads were difficult to spot while riding and caused us some consternation.

Unlike cacti, campgrounds were proving elusive in Arizona. Whenever we stopped for a break I would search for options on my phone but largely to no avail. That day I found what appeared to be a bargain motel, the Burro Jim in Aguila. The website was either off-putting or enticing, depending on your sense of humour. It proclaimed that thieves and pirates were welcome but truckers were not and everyone in between paid a premium. The language was bluer than the ocean and we couldn't fathom their target audience. Still, the reviews were decent enough, with one word standing out above all others: 'cheap'.

We rolled into Aguila at sunset and easily found the Burro Jim motel. It was hard to determine which building was reception so we knocked on the first door. The chap who answered was friendly but explained that unfortunately Burro Jim was no more so he couldn't rent us a room. Furthermore, he explained that regrettably the next motels were twenty-five miles away in Wickenburg. We apologised for interrupting his evening and resolved ourselves to another night on Highway 60. As we turned to leave the man beckoned after us saying that some rooms still had power and water and while he couldn't rent us a room, we were welcome to stay for free. Legend!

The room didn't have furniture but was still better than some we'd paid to stay in. We were on cloud nine. I walked to the nearest gas station to buy the man a six pack as a thank you for letting us stay. The attendant let out a bellowing "Howdy" as I entered and when I left he followed me out to the forecourt and asked "Which one's yours?"

"I didn't drive here" I replied.

Mind blown he bellowed back an elongated "You walkin' boy?" in a southern drawl, and then answered his own question with a long "Damn."

After my apparently unusual walk back to the motel I knocked on the man's door to give him his six pack. He politely declined saying he didn't drink and it was his pleasure, even checking with me that the power and water were working okay. I thanked him again profusely and returned to the room with an unexpected bounty of ice-cold beer. We laid on our foam camping mats in utter contentment, listening to music, playing gin rummy and wondering what Alexander Supertramp would make of our adventure.

CHAPTER TEN

DON'T MESS WITH TEXAS

3rd – 24th December 2016
Distance ridden this chapter: 1,124 miles / 1,809 kilometres
Total distance ridden: 6,657 miles /10,713 kilometres

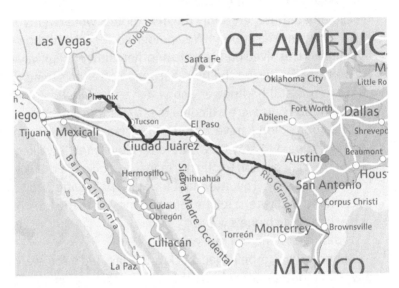

Imagine you're in a nineteenth century Wild West town. The local scoundrels are chained to a mesquite tree because the town doesn't have a jail. The saloons are brimming with likely lads spinning entertaining stories about panning for gold in the local river. Horses are neighing from the hitching rails and corrals. Chances are you're imagining something close to Wickenburg, self-proclaimed 'Dude Ranch Capital of the World' and originator of the term *Hassayamper*.

In the mid-nineteenth century, tales of successful gold panning on the Hassayampa river in Wickenburg were so rife that people joked you'd become a compulsive liar if you drank from its waters. Or as gold panners in the US more cutely put it, you'd become a *Hassayamper*. After our wonderous night in the defunct Burro Jim, we stopped at Wickenburg to make a coffee and take the obligatory photo next to the Jail Tree.

Having spent four of the previous six nights wild camping, we were fortunate enough to spend seven of the next eight nights with wonderful Warmshowers hosts. When we started using the service we had no idea how prevalent it would be and throughout the rest of the trip we were amazed at the remoteness of some of the hosts.

Our first host in Arizona was Bill, who lived in Sun City West. The contrast between Wickenburg and Sun City West couldn't have been starker. Wickenburg is all tumbleweed and authentic western 'howdy pard'ner', whereas Sun City West is your archetype artists impression of utopia. Built around ten golf courses, it's very *Truman Show*, and hardly believable as a real place. Bill was unnervingly generous, giving up his bed, taking us for a meal and, most excitingly for Ties, taking us to Ties' favourite second-hand clothing mecca Goodwill so we could replenish some of our hard worn threads.

From Bill's we headed through Phoenix via the city's canals and the exquisite mansion of chewing gum overlord William Wrigley Jr. We had a mission in Phoenix. Ties' inflatable mattress had been deflating frequently and it needed fixing pronto. It wasn't hard to find a camping megastore in the fifth biggest city in the US and the attendant quickly found a repair kit for the mattress. I was asking how to use it when news of our trip started spreading and another attendant said they'd fix it for me. After a bit of wrestling with the mattress they took it out back to immerse it in water, like you would an inner tube.

Sometime later they re-emerged, not with our mattress but with a brand new one.

"I couldn't find the puncture" the attendant said "and if I can't find it you certainly won't."

The attendant had called the mattress manufacturer and the company had told them to give us a new one. I was even given a refund for the patch kit I didn't need.

Our second hosts in Arizona were Mike and Jackie, both experienced cycle tourers and keen environmentalists. We had a fascinating tour of their garden that housed a homemade solar power and aquaponic system. This was another great thing about being hosted: before our stay with Mike and Jackie we had no idea what an aquaponic system was and now we were being shown first hand by resourceful and passionate environmentalists who mere hours beforehand we hadn't even known. Opting to use Warmshowers didn't just mean a comfy bed and a good scrub, it took us to unexpected places, taught us new things and introduced us to people that we're still in touch with today.

***

We zigzagged out of Phoenix and rode southeast to our next host in a tiny city called Coolidge, named after the President of the same name (Calvin Coolidge) who was in office when the town was founded in 1925. Our host Larry said he wouldn't be there but his brother and niece would be happy to host us. This was the first time we'd been hosted by someone who wasn't at home and again we couldn't believe the generosity of total strangers.

From Coolidge we continued southeast on the straight-as-an-arrow Pinal Pioneer Parkway, which kept us entertained with direct views of the Tortilla Mountains to the east. After thirty miles we stopped for lunch at the Tom Mix Memorial, placed at the site where the popular western actor died in a high-speed car crash in 1940. From the Tom Mix Memorial onwards we could see our goal for the day another thirty miles away; Mount Lemmon. The imposing mountain is a spectacular sight and we were fortunate enough to secure a tent pitch at Catalina State Park at the foot of the mountain.

After a short ride to the Rillito River, we joined the much hyped and wonderfully named Chuck Huckelberry Loop bike path that encircles Tucson. Aside from the highly recommended bike path, we had no expectations about Tucson. We didn't know anything about it, but it would go on to become one of our favourite cities of the trip, and certainly the most cycle friendly in the US.

The Chuck Huckelberry Loop lived up to the hype, with over 130 miles of paved bike path, along with useful amenities and loads of interesting sculpture art. There was a bizarre bats and bikes theme to some of the sculptures, including a depiction of two bats riding bikes high in the sky, titled Extreme Batty Bikers, and two bats on a tandem towing their baby bat, called Batty Biker Family.

Tucson was also home to our next hosts, Randy and Cheri. Randy is a master jewellery maker and Cheri a coloratura soprano. It didn't take long after we arrived before Cheri burst into song and as soon as we were settled, Randy suggested we take a short bike ride to their favourite Mexican restaurant for a late lunch. It would turn out to be a highly entertaining ride following our hosts' tandem through Tucson. Randy was upfront piloting the machine whilst Cheri, resplendent in her fluorescent tie-dye shirt, was belting out operatic melodies at decibel levels I'd not believed human beings capable of producing.

The next morning Randy fuelled us up on his mouthwatering chickpea pancakes smothered in his homemade prickly pear syrup. Over breakfast, Randy told us about the Davis–Monthan Air Force Base – the world's largest aircraft boneyard – and offered to escort us there on the Chuck Huckelberry Loop. The boneyard was a phenomenal site, housing 4,000 dormant military aircraft and intercontinental ballistic missiles, all painstakingly lined up in neat rows.

After a week in Arizona, we were becoming accustomed to the surprisingly hilly landscape. However, we didn't expect Arizona to take us higher than the Rockies. As we crested the Mule Pass, we exceeded 6,000 feet for the only time on our trip. Our reward was a hair-raising descent into the old copper mining town of Bisbee, where we stayed with our friendly host

Stephen. Stephen was keen to show us a good time at Bisbee's fun bars, but after reaching the heady height of 6,000ft we could barely muster the energy to blow up our air mattresses. Sorry, Stephen!

On our last day in Arizona, we spent fifty miles riding alongside the Chiricahua Mountains and hardly saw a single vehicle. It was a fitting end to what was at that point our favourite state and one we'd recommend to anyone whether cycling or not.

***

Not long after crossing the border into New Mexico we entered the community of Rodeo. With a population of 101 it's tiny, but still has shops, a museum, a post office and an airport. It also has our favourite campground in the US, Rusty's RV Ranch. We chose to stay at Rusty's for purely practical reasons. We were about to ride along Highway 9, which is largely bereft of campgrounds. It also hugs the Mexican border and as such is one of the most heavily patrolled roads in the US. This ruled out wild camping; we didn't fancy being woken up for a grilling by US Border Patrol Agents.

We had no idea that Rusty's was renowned as a world class astronomy site and home to some of the darkest skies in the world. The campground caters to enormous rigs, some of which had their own observatories attached or on tow. One RV owner excitedly told us that a there was a perigee-syzygy (a 'supermoon') the following evening that would light the whole area up in the middle of the night. After hearing this – and seeing the seventeen-foot hot tub on offer – we decided to splash out an extra $15 and have a rest day.

A supermoon is a full moon that is particularly close to Earth and therefore appears around thirty percent brighter than a normal full moon. Fortunately for us, the supermoon did appear – one of three that materialized in 2016 – and it was one of the most extraordinary natural wonders either of us have ever been privileged to see. As we wandered around the campground it was hard to believe it was 3am and not midday, such was the impact of the intense yellow glow.

The New Mexico desert lit up by a supermoon at 3am

Well rested, we left Rusty's the following day with a strong tailwind and smiles on our faces. Highway 9 was deserted, and the few vehicles we did see tended to be US Border Patrol Agents. It took us two days to ride the 164 miles from Rusty's to El Paso. In that time, we lost count of the number of US Border Patrol Agents who pulled alongside us in their vehicles to check us out. After the first couple of interrogations, we realised they just wanted to hear our accents. I hammed up my Queen's English which seemed to placate the officials.

In addition to the patrol cars there are also frequent CCTV cameras, drones and helicopters. After being followed by helicopters on several occasions, we finally released that if we waved at them this seemed to satisfy the agents and they flew off. All the security seemed a bit pointless to us but was, we supposed, an inevitability in the new Trump-era.

The desert along Highway 9 is desperately barren but we were told this is fertile roadrunner territory. We managed one sighting of the awkward creature when a bird of prey swooped down into some scrub next to us. A roadrunner scurried out of the bush and under the safety of some boulders. Watching the scene from our bikes was surreal, and it really felt like David Attenborough might emerge at any moment.

***

We had been given more warnings about road safety in El Paso than in any other city in the US. Luckily one of our Instagram followers James sent us detailed instructions for navigating the city sprawl. With James' route guidance we left El Paso with a rosy impression of the city and spent the rest of the day riding through countless pecan farms.

At Fort Hancock we had to re-join Interstate-10 as our only option for the sixty-seven miles to Van Horn. Ties has fond memories of this day. Not because of the landscape or people we met, but because of the glorious thirty to fifty miles per hour tailwind that pushed us all the way. At its most extreme, the wind was assisting so much that we were freewheeling our heavy loads up hills with seven percent gradients while cars and trucks honked and waved on the highway sharing our high-speed freewheeling joy. We even picked up a Strava trophy that day, as the seventh fastest people ever to ride a seventeen-mile section of Interstate-10. Admittedly there aren't many others on the leaderboard but our time that day is, as I write, still good for fifteenth out of fifty-six people to have completed the segment.

We felt blessed, but also somewhat deserving of the assistance after the drudging headwinds we had endured four months earlier in the Prairies. Not only did we benefit from strong tailwinds that day, we also basked in the mid-December twenty-four degrees Celsius (seventy-five Fahrenheit) heat. Ties was on cloud nine as we rolled into Van Horn and was excited to know where we were camping. She was rather taken aback, but pleasantly surprised, when I announced that I'd booked a Motel 6.

"Awesome, but why?" said Ties "the weather's amazing, are we celebrating something?"

I explained that I was finding it hard to believe but the forecast said the temperature was due to drop from the current twenty-four degrees Celsius to minus fourteen degrees Celsius overnight.

True enough, as we left Van Horn the next morning it was minus-eleven degrees Celsius with a windchill of minus twenty (minus four Fahrenheit). We didn't have appropriate kit to camp in these temperatures, so had no choice but to ride seventy-five miles to the town of Marfa into a biting headwind. To spice things up a bit Ties' rear wheel punctured thirty minutes into our ride. Changing an inner tube on a touring bike is never fun, but at minus twenty it's the polar (pun intended) opposite. As I was fixing Ties' puncture, she noticed something coming out of her water bottle. After thirty minutes on the road our water had frozen solid and forced its way out as ice.

Frozen water bottle within thirty minutes at minus eleven degrees

Celsius

We at least experienced some light relief when we stopped to make lunch at the Prada store in Valentine. Although it contains actual Prada handbags and shoes, the tiny building is not a real store but an art installation. The idea was to create a biodegradable structure that would slowly be consumed by the elements; an ironic take on the material consumerism of luxury brands. Whatever your take on the concept it's fun to see and a cycle tourers delight, providing us with shelter from the wind and a handy outside power outlet.

As we rolled into Marfa, we finished the day in darkness for only the second time on the trip. We were disappointed not to see more of Marfa – as we'd been told it was an interesting town – but the freezing headwinds had made progress slow.

I'd found a room on Airbnb that was in a house converted from an old train depot. The owner had filled the enormous space with eccentric art and it was a feast for the eyes after ten hours in the desert. After installing in our room, we headed to the supermarket but it had just shut for the day and we ended up with our one and only Dairy Queen burger of the trip. A truly underwhelming experience.

After our long days in New Mexico and the gruelling ride to Marfa, we desperately wanted a rest day. However, we couldn't afford another night in the fancy artist's residence and were on a schedule to reach Ties' friend Aaron – who she met in Mexico at one of her Meetup events – in San Antonio for Christmas. We decided to have a short ride to the ironically named desert town of Alpine, where we were confident of finding a cheap motel.

We were now deep into the Texas Mountain Trail Region and enjoying the majestic scenery. Whilst the elevation of the road remained steady, we were surrounded by mountains that looked all the more impressive protruding from the dusty grasslands.

Alpine delivered exactly what we needed and we had a fun afternoon checking out the Christmas decorations and public art. The next day we passed a fake Target (sort of like an

American Debenhams) store, fashioned out of an abandoned railway building in the middle of nowhere. It seemed like a rustic version of the Prada store we'd seen two days ago and as with its fancier cousin we stopped to have lunch. Sadly the fake Target store is no longer there, as the owner became jittery that the decrepit building might collapse on people like us, and tore it down in December 2020.

The 'Target store' before it was torn down

We called it a day not long after seeing the tiny Target store, at a campground in the small town of Marathon. As we were setting up, fellow travellers Bill and Colleen came over and after a quick chat invited us for dinner at their RV.

We asked how we'd find the RV and Colleen said "Easy, that one", pointing at what looked like the land yacht from the 2004 film *Meet the Fockers*.

This was exciting. We'd stayed in over eighty campgrounds

by this point but never had the opportunity to see inside one of these beasts.

After a good scrub we put our best threads on and strolled over to the RV. Bill answered and insisted on showing us the outdoor TV before we entered. Surely not! With the touch of a button, a hitherto innocuous panel cantilevered out, flipped, and exposed a sizeable flat screen TV. After a tour of the luxurious interior, we took our place at the palatial dining table. Bill asked if burgers were okay and then proceeded to make meat patties from scratch in the well-equipped kitchen.

The next morning, we bumped into Bill and he beckoned us over to show us a special plant. Bill explained that the lofty Agave Americana we were looking at flowers once in its life and then dies, which is why they're commonly called century plants. This one was definitely in bloom and we felt both privileged to witness this event and grateful for Bill's knowledge and insight. As we left the campground, Bill and Colleen wished us bon voyage and filled every nook and cranny of our panniers with cereals bars and dried fruit.

***

Cycle touring down the Pacific Coast is so popular it almost feels a bit touristy. If that's not your vibe, then the Southern Tier is the complete opposite. Two days after leaving Marathon we saw our first cycle tourers since California. We were ascending when we noticed what we thought were people at the top of the hill. As we drew closer, we realised they were cyclists and closer still we realised one was taking our picture.

The exceptional photographer was Ryan, a nurse from Virginia, and his friend was Nick, a teacher from New York City. Ryan and Nick were heading west, with no specific agenda or goal other than to have fun. We exchanged details and Nick said if we made it to NYC we could stay with him. Months later we would take him up on the offer. And years later Ryan would stay with us in Cornwall!

Don't mess with Texas

We had planned to end that day at Langtry, which is only separated from Mexico by the Rio Grande and the closest we would come to the country where Ties and I met. The campground we planned on staying at had seen better days but it was worth the stop as we were given some 'Don't mess with Texas' bumper stickers from the visitor centre.

Although the phrase originates from a 1980's litter campaign, it'll be forever associated with a 2001 *Saturday Night Live* sketch. Will Ferrell, as George Dubya, tells a terrorist "You violated rule numero uno. You messed with Texas. Don't mess with Texas."

Ties still has the sticker on her mudguard to this day. And it still regularly raises an eyebrow.

Armed with our stickers, we plodded on another twenty miles to make it an eighty-one-mile day and end at Seminole Canyon State Park, the best US State Park we camped at, and the cheapest campground of the whole trip at the princely sum of $8. The showers were so powerful at Seminole I didn't want to leave and couldn't believe they weren't tokened. The only thing Seminole didn't have was bear lockers.

By this point I had become a dab hand at making a bear locker, perfecting the rope-over-the-branch-as-winch technique. However, there aren't exactly many trees in the desert. Our back up was to simply use our bike lock to tie the panniers to something like a picnic table. Not exactly bearproof, but it kept us safe in the tent.

At Seminole I was woken up in the middle of the night by rummaging. It sounded like a small animal but I was wary of the warning signs dotted around for bears and mountain lions. I opened the zip on the tent painstakingly slowly as to not startle whatever critter was out there. In the pitch black of the desert, I could only make out the silhouette of the animal. Importantly, I could tell it was indeed small. And certainly not a mountain lion. I put my headtorch on and as two tiny eyes stared back at me I could see it was an armadillo, which promptly pegged it at lightning speed. I retreated into the tent, confident I wasn't going to be eaten by the armoured fella and pleased I hadn't scared the armadillo into its defensive bowling bowl shape.

***

We told Aaron that we'd arrive in San Antonio on Christmas Day. We were on track but Aaron and his friends were having a party on Christmas Eve and they wanted to meet us. Aaron also insisted that Highway 90 – our intended route into the city – was unfit for cycling. Aaron was not going to back down and insisted on picking us up from wherever we could ride to by lunchtime on Christmas Eve.

Ties was excited about seeing Aaron and the prospect of a party with new friends. This will sound ungrateful, but I had mixed feelings. I appreciated Aaron's offer but I had a needless bloody-minded determination to cycle the whole way. I had also been using an online plug-in built by a community Strava user called Jonathan O'Keeffe to build a map of our route. I knew Highway 90 was a dead straight road and would show up as annoyingly big gap in the map. Eventually, sense prevailed. I got over myself and we arranged to meet Aaron at Walmart in Uvalde.

By the time we rolled into Walmart's car park we'd ridden eleven days without a rest day, clocking up just short of 700 miles in the process. We propped our bikes up on a lamppost on a grassy verge, rolled out our foam mats and sat down to make lunch. Ties goaded me about my shaggy bouffant and Brian Blessed facial hair. I was proud of the beard I was cultivating, but Ties insisted I was to have a Christmas trim at Aaron's.

As I tucked into my peanut butter sarnies, a lady came over with her young daughter. We said 'Merry Christmas' and without responding to us we saw her take out some money and give it to her daughter to give to us. This was awkward. We'd had some generous donations from strangers before, but only after we'd chatted about our trip and people had said they wanted to contribute to our adventure.

This lady clearly thought these two bedraggled people eating a limp sandwich in a Walmart car park on Christmas Eve were homeless.

"Thank you, but we're not homeless" I said "we're cycle touring and waiting for a friend to pick us up."

The lady turned beetroot red. Ties was stifling laughter. I went crimson under my Brian Blessed disguise.

The lady uncomfortably retrieved the cash from her daughter's hand and sheepishly moonwalked away from us. I turned to Ties and said "okay, I'll have a trim at Aarons."

It didn't register, Ties couldn't stifle her laughter any longer and was in raptures on her foam mat.

Attracting flies (look closely at my back) on Christmas Eve in Walmart car park

CHAPTER ELEVEN

COMMANDER-IN-GRIEF

25th December 2016 – 22nd January 2017
Distance ridden this chapter: 1,209 miles / 1,946 kilometres
Total distance ridden: 7,866 miles / 12,659 kilometres

Our Christmas in San Antonio was special. We spent three days hanging out with Aaron and his friends, never once feeling like interlopers. On Christmas Day itself we ambled along the city's river and had a Tex-Mex feast at one of the many waterfront eateries. We visited the Alamo and had our picture taken with the rangers who protect the landmark to this day. In the evening we visited Aaron's friends and had our second Christmas Day Tex-Mex feast, all washed down with plenty of cerveza and tequila.

On the day we left San Antonio, Aaron spent the morning driving us all over the city so we could replenish our provisions and acquire some essential new cycling kit. As he did so, Aaron explained San Antonio's beguiling road network to us.

Like London, San Antonio is encircled by a series of concentric loop roads. Being the US, San Antonio's are bigger, and they have one more than London. The second Loop '1604' is similar in size to London's M25 and the third loop is over 200 miles in circumference.

After ferrying us around the city, Aaron insisted on dropping us at New Braunfels to transport us outside the 'outer-outer loop'. I wasn't going to argue, grateful for Aaron's unwavering hospitality and accepting the gap in our map was going to grow bigger.

From New Braunfels we were finally able to remount our bikes and tackle the short ride to San Marcos, where we had a fun night staying with our hosts Kerry and Alyssa. Kerry is an exceptional story teller and had us in stitches all night, regaling us with tales of his cycling exploits and his many talents as an inventor, photo taker and music maker.

A few miles into the following day's ride a Texas State Trooper drove past us. Ties is much more gregarious than I am and loves to smile and wave at people, including law enforcers. Normally, when Ties excitedly waved at the Police, they carried on. Maybe they didn't even notice. That day, the Trooper turned his siren on, did an impressive about-turn in his Texas State Trooper-liveried Dodge Challenger, and stopped abruptly in front us.

The smile instantly disappeared from Ties' face. We both stood there, frozen to the spot for what felt like hours before the trooper finally emerged from his raucous vehicle. He sauntered over to us, stereotypically flicking through his diminutive notebook as he did. Then he politely greeted us and carried on to explain that he was about to take on the southern section of the Great Divide bikepacking route, and he'd love to pick our brains.

We must have both exuded palpable relief, as he added "Gosh, y'all dint think y'alls in trouble did ya?"

"Kinda" I snapped, in an involuntarily irritated tone.

We all had a rather awkward laugh and we gave the Trooper some tips. Then he sauntered back to his Challenger and sped off with what sounded like an extreme amount of horsepower.

Relieved from our friendly encounter with the local Texas plod, we kept going until Austin where we were staying with Jen and her family. Jen is a friend of Jess, who we stayed with in Renton. She and her family live in the West Hills area of Austin, where we experienced our first ascents of the trip over twenty percent. We had an entertaining rest day hanging out with Jen's family and then headed east for a few mercifully flat days through Texas farmland.

***

Before Ties left Mexico, she spent one of her last nights watching baby turtles hatch at Playa Akumal with Texan sisters Linda and Joyce. Ties stayed in touch with the siblings throughout the intervening years and when we reached Conroe, we met up with them at the old animal hospital owned by Linda's husband. Conroe is a small city directly north of Houston, whose 1930s oil boom meant it once boasted more millionaires per capita than any other US city. There was no sign of such lavish wealth when we rode through the centre but we had a lovely evening chatting to Linda and Joyce, whilst devouring a massive Tex-Mex meal. After dinner we had the privilege of staying in the old animal hospital, which was still home to two playful kittens that found our bikes and gear fascinating.

From Conroe the terrain was so flat we rode eighty-eight miles to the small town of Silsbee. We found a super cheap campground and rewarded our day's effort with a taste of Americana from the 1950's diner on site.

Our final day in the Lone Star State rounded off our positive experience of Texans. Early on in the day we pulled into a gas station to buy some provisions and met a delivery driver. Ruby had a packed schedule but she took time to hear our story and hand us some cold bottles of water from her cab. Later that morning we saw Ruby again and she offered to take us for lunch at a local Mexican restaurant. We met her there and she treated us to a delicious meal of authentic Mexican food.

After lunch we said our goodbyes to Ruby and an hour later she stopped in front of us again. Ruby explained that the next ten miles of road were drab with no shoulder, and asked "would we like a lift."

Without giving us chance to answer she dropped the tailgate, we loaded our bikes into the back and Ruby jettisoned us to the Louisiana border.

***

Ties is a spiritual soul and was excited about the aura from Louisiana's bayous and rich Creole heritage. She had also been dreading the start of our Deep South stretch for one reason; dogs. We both love dogs but the Deep South is infamous amongst cycle tourers for its abundance of freely-roaming, cyclist-hating dogs. Ever since the Pacific Coast, Ties had been making preparations by asking every cycle tourer we met how they deal with the issue. Their answers were hugely varied. We met plenty who strapped a stick to their top tube and would lash out near the dog to deter them. The most laid-back approach was from an Amish cycle tourer we met in California, who swore by the technique of calmly talking to canines. Others were less understanding, liberally unloading pepper spray whenever a dog came tearing after them.

It didn't take long before we had to decide what our technique would be. On our first full day in Louisiana we were chased by countless dogs, all seemingly dead set on sinking their teeth into us. Ties and I had opposing theories about how to deal with this. My approach was simple. Pedal fast, really fast! Ties thought this would provoke the dogs into a chase and she chose to simply freeze on the spot when dogs launched an attack. For the first few days I would instinctively sprint away from the dogs and invariably have to backtrack to rescue Ties from the pack of baying hounds that had encircled her. Eventually I persuaded Ties that flight was better than fright and she joined me in what she would come to call our hourly 'spinning class'.

Normally my technique worked. The dogs were not stray, they were guard dogs that people let lie untethered in their front gardens. When we rode by, the dogs would instinctively jump up to defend their territory and would not stop until we were several plots of land away whereupon they would turn back home, satisfied that another enemy had been vanquished and a job had been well done. Sometimes it was almost funny. On one occasion a little terrier latched onto the heel of my shoe. Try as I might no amount of jerking would shake the critter loose. Eventually I had to stop riding and pull the terrier off, who looked discombobulated and forlornly set off to find home.

At other times it was terrifying, with packs of Pit Bulls, German Shepherds, Dobermans and Rottweilers launching after us. On one such occasion we were surrounded by frothing dogs that forced us into the middle of the road. Several drivers, including an articulated lorry driver, spotted what was going on and used their vehicles to manoeuvre between us and the dogs. It was amazing how these drivers choreographed their vehicles to gently nudge back and forth, stopping the dogs coming our way for long enough for us to scarper. This was, I thought, nobody's first rodeo but mine.

The dog chasing was not just worrying from our perspective, there was an even more upsetting side that couldn't go ignored. From the moment we entered Louisiana we saw dead dogs on the side of the road every day for the next 3,000 miles. It was so heartbreaking to see the dead pets which could have been so easily saved if only the dog owners hadn't let their dogs roam free. We couldn't understand why this was so common in the south-eastern states, but our experience sadly backed up the warnings from the other cycle tourers we'd met.

***

Other than the dogs something else followed us into Louisiana; *Storm Helena*. This was the second cold snap of the trip and our first snow, not something we expected in Louisiana. Luckily we found several friendly Warmshowers hosts who sheltered us from the weather. We spent our first night in the state with Taylore and Luke. Taylore is an experienced cycle tourer and told us all about her trans-America tour from Portland (Maine) to Portland (Oregon), while Luke made us a delicious dinner.

The snow came the following day as we arrived at our host Jordy's house, near the small city of Eunice. Cajun culture is important to Eunice, a city that founded the Cajun Music Hall of Fame and Museum, hosts one of the largest Courir de Mardi Gras (traditional Cajun festival) and the World Championship Crawfish Étouffée (Cajun stew) Cook-Off. Jordy was a croupier at the local casino and as he worked the following day, he let us hibernate in his home for an extra day during the worst of the snow. During the evening he taught us all about Cajun and Creole culture, made us a mouthwatering venison sausage jambalaya and introduced us to boudin; a blood sausage with the added bonus of rice and veg mixed in with the pork scraps. Boudin became our staple lunch as there was always a local vendor parked up at gas stations in Louisiana.

Our next host, Ruth, lived in a historic pecan plantation. The house was filled with curiosities and decked out in neo-classical style. Ruth offered us tea – as one should have in a plantation home – and then suggested we go to a local eatery for dinner. I took the opportunity to order the most Louisianian of Louisiana dishes, a blackened catfish po' boy. Louisianians love po' boys so much there are huge festivals dedicated to the dish. Before I met Ties she attended a po' boy festival and was enthusiastic for me to try one. For the life of me I can't understand the excitement. I love a good sandwich, but that's all a po' boy is.

To this day Ties insists it's different because it's a Louisianian sandwich. We've agreed to disagree.

Louisiana also unnerved us with some awkward moments being hosted by open Trump supporters. We had a fun evening with one seemingly liberal couple, Carl and Marilyn, who then proceeded to proclaim the next morning that president elect Trump would be the US's first great leader since Reagan. This was extremely awkward. Our host's statement came from nowhere. Silence descended for a few seconds before Ties couldn't help herself and launched into a tirade about the many political and personal achievements of Obama. Carl was unconvinced, believing that Obama was far too focused on ethnic minorities and created resentment amongst neglected, white working-class Americans. It was a bizarre outburst that made for an uncomfortable departure from an otherwise friendly couple. After this altercation, we made a conscious decision not to engage in political discussions with our hosts and just enjoy the social contact regardless of political views.

That evening we stayed with another friendly host, Shirley, who went out of her way to make us homemade Louisianan fare. As we relaxed in our host's lounge, Meryl Streep was using her Golden Globe speech to give a scathing indictment of Trump. Jeff Bridges admired the speech so much that he knelt down next to Streep, striking a full-on *Wayne's World* 'we're not worthy'.

Our host wasn't as enthusiastic, spitting feathers about the audacity of this woman – "who isn't even American" – to criticise Trump.

Ties and I looked at each other quizzically and both hurriedly checked our phones. Yup, thought so, Meryl Streep is indeed American. While I mulled this over in my head I threw Ties a hardy stare that she knew meant 'remember what we agreed this morning; don't rise to the bating'. Ties bit her lip and we had a civil evening while Shirley told us some truly bizarre stories about her penchant for bizarre fetish events that cannot be retold in this book.

***

For me, the jewel in Louisiana's crown is undoubtedly New Orleans (NOLA). We used the generous Christmas money gifted to us from friends and family to spend a few days in a B&B not far from NOLA's French Quarter. Ties loves the architecture in NOLA and we spent hours aimlessly wandering the streets admiring the creole style and rummaging through the boutiques. We found one shop that made remarkable 3D laser cut maps from wood. Two years after our return to the UK, Ties contacted the shop to order me a 3D laser cutting of the Great Lakes for my fortieth birthday. This is the perfect memento from our trip and takes pride of place on our living room wall in Bristol.

NOLA strengthened our love of Louisianan fare. We dined at a local Southern eatery near our B&B called Horn's and were introduced to the delights of pulled pork, pickled peppers and chimichurri on a cornbread waffle for breakfast, or 'waffle cochon' as Horn's call it. We indulged in many beignets – posh doughnuts – at the famous Cafe Du Monde. We took a bus ride to imbibe local craft beers at the inventively titled New Orleans Lager and Ale (NOLA) Brewing Company. And of course we found our way into some of the tiny jazz bars of the French Quarter, avoiding the tacky tourist haven that is Bourbon Street.

From NOLA we entered Mississippi crossing the East Pearl River. As we navigated our way to the campground at Buccaneer State Park, we saw the ocean (Gulf of Mexico) for the first time since San Diego. This was also the start of our journey along the Gulf Shore, a stunning shoreline that stretches from Mississippi to Florida, characterised by calm blue sea and powdery white sand beaches. I knew nothing of this part of the US and was amazed by the endless beauty of the coastal landscape and relaxed vibes of the whole area.

As we cruised along the bike path that hugs the Gulf Shore,

we crossed paths with the most heavily-laden cycle tourer we've ever seen. Thomas was riding across the US with his chiweenie – a cross between a Chihuahua and a miniature Dachshund – Charli, perched in a plastic crate strapped to the top of his front pannier rack. Charli looked like a ship's figurehead, with her snout protruding proudly from the sanctuary of her crate. The amount of stuff Thomas was carrying was extraordinary and it made for easy conversation that went in an unexpected direction.

After the usual cycle touring protocol of discussing routes and recommended campgrounds, Thomas asked "what do you carry?"

I naively responded "our bikes weigh fifty-five-kilograms, but you must be carrying way more than that?"

Thomas looked confused and replied "no, what do you carry?"

Thinking he needed an imperial conversion, I said "oh, our bikes each weigh about 120 pounds."

Frustrated, Thomas leaned in towards us and asked "what guns you got?"

Much to Thomas' amazement, we explained that we didn't have any guns and it hadn't even crossed our minds to burden ourselves further with superfluous weaponry. Thomas thought this was foolish and talked us through his trio of guns that were strategically stashed in his panniers, close to hand. He told us how he wouldn't go anywhere without a gun and then set off on his steed with Charli's nose pointing due west. This wasn't the last time someone asked us 'what we *carry*' and our encounter with Thomas ensured we didn't respond as naively as we did that day.

***

Whilst the Gulf Shore was giving off tranquil vibes, the Department for Homeland Security was doing its best to stress us out. US visa rules stipulate that Homeland Security has until

Commander-in-Grief

seven calendar days before the expiry date on an I-94 visa extension request form to give hopeful travellers the 'yes or no' verdict on their visa extension application. Our expiry date was 15th January 2017, the day we left Mississippi. The 8th January came and went and we hadn't heard from Homeland Security, so I called them. I was told "not to worry about it" because they were over thirty days behind with application processing.

This wasn't a particularly reassuring response, but we didn't have many other options so we carried on riding.

Our visa status was the cause of some consternation. We were about to become illegal aliens in the US and in five days' time Trump was being inaugurated as President and *commander-in-grief*. No one knew what this would mean exactly, but all I knew was I didn't appreciate the uncertainty. We started to assess our options. We considered heading back to Toronto, but it would have been far too cold there in January to continue cycle touring. For the first time in over fifty years, direct flights from the US to Cuba had been authorised in June 2016, and we came close to heading to Havana. However, that would have meant passing through US customs with expired visas so we ruled that out. We even seriously contemplated cancelling the whole trip and heading home early. Eventually we decided to take the risk and follow Homeland Security's advice 'not to worry about it'.

After two more days riding along the glorious Gulf Shore, we stayed with Ties' friend Ginger in Elberta, Alabama. Like many Americans, Ginger's partner found me difficult to understand. It had become a running joke that Ties often had to translate for me, with most Americans finding my 'Australian' accent too hard to comprehend. Ginger's partner was hard of hearing and this exacerbated things. After repeatedly asking me to speak up he eventually became so frustrated he pulled me in towards him, nose-to-nose, and bellowed "talk to me like a man."

Conversation was easy after that, although the next day I was a little hoarse after spending all evening yelling at Ginger's partner. Ties found this hilarious and has never let me live it down.

The following day we entered Florida. Over the next month we would go on to ride 1,500 miles in Florida, narrowly beating California as our most travelled state. We entered Florida over the Perdido Bay Bridge and quickly made our way to Santa Rosa Island. This stretch of the Gulf Islands National Seashore is breathtakingly beautiful. Other than a small area at the start of the island, it's a rare stretch of completely undeveloped Florida shoreline. To top it off, there is a bike path that runs parallel to the one road along the island, giving unobscured views of the white sand beaches and turquoise ocean.

The remainder of the Panhandle coast is just as attractive, albeit decidedly more developed. At Port St Joe we stayed with Ted in his luxurious shorefront home. Ted's house is in a gated development and seconds after entering we were intercepted by the armed private security guard, in his unnecessarily big patrol car. He stepped out of his car and, without a hint of irony, put his hand on his belted gun and extended the other arm towards us palm up. We stopped and he barked "You can't cycle through here."

We explained what we were doing and he only let us carry on once we showed him our Warmshowers message history that included Ted's address.

Ted was far more laid-back than the security guard, finding the story both surprising and funny. We had the pleasure of staying in Ted's carriage house, a self-contained suite above the detached garage, connected to the main house via an elevated walkway. After a good wash, we joined Ted in the main house and spent the evening dining and drinking while taking in the ocean view through the huge panoramic glass walls. When we left, Ted gave us his details and said if we time our transit through Maine we might catch him at his other house and would be welcome to stay. Three months later we were fortunate enough to enjoy Ted's company again.

***

On 20th January 2017, as Trump was officially introducing Voldemortian politics to the US, we rode along the Big Bend Scenic Byway to Eastpoint. We had rented an Airbnb room and were looking forward to a rest day. Our host Sharon was preparing for her own adventure the next day, as she joined one of 680 Women's Marches challenging Trump's objectification of women, and promoting human rights worldwide.

Sharon let us borrow her car and we used it to visit the local nature centre. When Sharon returned from the march she took us for dinner and then gave us a tour of the stunning St. George Island.

We left Sharon's home in the midst of some ominous looking weather but somehow managed to stay dry all day. As we rode through the tiny community of Medart we passed a Trump / Pence sign that someone had doctored with a banner saying 'ALL HAIL'. Great, we thought as we looked for a campground to end the day, considering what sort of people might be hailing the new president. A few miles later a car pulled alongside us and told us to leave the road as there were multiple tornado warnings. Great, we thought again, having déjà vu from Manitoba, albeit this time there was no heroic Morton or nearby motel to hide in.

We pulled into Newport County Campground where we were surprised to hear loud music and see lots of dancing. Half the campground seemed to be taking part in a tornado party and the other half seemed to be hurriedly packing up all their stuff and hightailing it. We booked a pitch but didn't set up, opting to rest our bikes against the washroom building where a handful of people were congregating.

Within minutes we noticed a burly bloke strategically repositioning his massive café racer motorbike. He then casually sauntered over and said to us "I wouldn't stand there, it's not structural."

He explained that all the people hiding in the washroom would be toast if the tornado came and we were better off joining him in the open sided shelter opposite. I didn't understand his logic, but his Steve McQueen stature and confidence were much more reassuring than the twitchy folk milling around the washroom.

We sat with 'McQueen', as I dubbed him, for hours, drenched by the horizontal rain barrelling through the open sides of the shelter. I must admit that this was the only time of the entire trip I was genuinely scared. For some reason the Prairies tornado and Oregon typhoon hadn't troubled me too much. The dogs had irritated me but I didn't find them overly irksome. But this tornado was really giving me the willies.

I asked McQueen what we'd do if it arrived. He showed me his tornado tracking App, demonstrating we were right in its path and then said "if it continues on this path, we'll leave the shelter and lie on the floor."

This was real squeaky bum territory and I couldn't understand how Ties was so unfazed. I reasoned that McQueen must be giving her more reassurance than I and resolved to put more faith in our companion.

Half an hour later, with the tornado party still in full swing, McQueen casually strolled over to his motorbike and moved it over to his pitch. I rushed over to McQueen and blurted out "Where are you going?"

He explained that the trajectory had changed and he was going to return to fiddling with his iron horse. He followed this up by adding that there was now a flood warning and he wouldn't like to be camping. This was unwelcome news but it was now pitch black and we weren't about to start riding again in an area where the local residents were hailing Voldemort and the storm was still brewing. We put the tent up and retired to bed, hoping we didn't wake up floating again.

I hardly slept all night, terrified of the flood warning and worried that every snapping branch signified the arrival of a tornado. Frustratingly unrested but nonetheless pleased to be dry and intact, we emerged from our tent early next morning and left the campground extra-early. As we rolled down the access road, we saw the worryingly high watermark on the trees that lined the road and realised the prominent position of the campground had saved us from the flood.

We were lucky on that day in Florida. Between 21st-23rd January 2017, twenty people were tragically killed by eighty-one tornadoes in the South-Eastern US. In hindsight we ought to have turned around when we were first warned about the tornadoes and gone back to the nearest town to pay whatever we needed to stay indoors. Hindsight is a wonderful thing however and I resolved to pay much closer attention to the weather forecast for the remainder of our trip.

CHAPTER TWELVE

SEE YOU LATER ALLIGATOR

23rd January – 7th February 2017
Distance ridden this chapter: 724 miles / 1,165 kilometres
Total distance ridden: 8,590 miles / 13,824 kilometres

In the UK in the 1980's, Florida was considered *the* location to go on holiday. The rich kids at school would come back from summer holidays with tales of Mickey Mouse, Universal Studios and something called McDonalds. Kids waxed lyrical about the fast-food chain that was a bit like Wimpy, but served something called fries instead of chips and gave you your meal wrapped in paper, rather than on a china plate. That sounded unorthodox. In 1992, the first McDonald's opened its doors in Plymouth, and I eventually sampled my first Big Mac. I was disappointed, but the underwhelming meat patty had not taken away my appetite for the Sunshine State.

As I grew older the bronzed bodies and blue rinse reputation of Florida's coast lost its appeal somewhat. However, the Florida Keys continued to intrigue. Ever since Brian Wilson sang about the mythical island 'Kokomo' in 1988, and Arnie blew up the Seven Mile Bridge in the 1994 film *True Lies*, I had wanted to visit the utopic archipelago. However, even with this sense of anticipation, I had never expected Florida to be our favourite US state of the trip.

***

At Cross City, where Florida's panhandle joins the west coast, a passer-by pointed us in the direction of Florida's Greenways; an epic network of paved bike paths that we barely left for days. The path took us through forests, parks and wildlife reserves providing a feast for the eyes and easy pedalling. As the path snaked through Crews Lake Wilderness Park we bumped into Kathryn, a cycle tourer from the US who had been on the road for three years on her three-wheeled recumbent: a bike that places you in a low position close to the ground, with your legs stretched out in front. Kathryn had a long aerial on the back of her trike with flags from all the countries she'd ridden through, including New Zealand, Australia, Malaysia and Singapore, as well as numerous European countries.

Kathryn may have been a few years our senior but she was also undoubtedly much cheekier, taking a bullish approach to wild camping and air travel. In her three years on the road she'd hardly paid for a single nights' camping and never once incurred a fee for a bike on a flight, insisting it was a 'mobility aid'. We loved Kathryn's attitude and rode with her for an hour or so, taking turns on the recumbent that we would later discover was made one mile from our future home in Falmouth, Cornwall.

Kathryn on her well-travelled trike

We left Kathryn at a water fountain on the bike path, where she intended to spend the night, and headed west into the Jay B. Starkey Wilderness Park. It's impossible to overstate how beautiful Starkey Park is. The bike path through the park is less than ten miles long but I could eulogise for hours about its tranquillity, and the beauty of its flora and fauna. The landscape feels like a safari plain, chock full of vibrant wetland plants and trees, dominated by the impressive Pine Flatwoods and Cypress Domes. The swamps play a vital role in the parks' ecosystem, feeding a cocktail of minerals to the grasses and trees that reside in the water. As we rode through there were animals darting around everywhere. I almost hit a deer that bounded across the path in front of me – like a scene from Bambi. There were birds of all shapes, sizes and colours chirping away in the water and the low canopy. We saw plenty of the aptly named Little Blue Heron, as well as the colourful Wood Duck and noisy Hairy Woodpecker. It was perfect. I can't think of anywhere else I'd rather be.

A few miles after Starkey Park we met with Craig and Diane, who were to be our hosts for the evening. We had an entertaining evening, listening to stories from their cycle tour of New Zealand and being spoilt rotten with food and drink. Craig invited us to stay another day and took us kayaking on the Chassahowitzka River, a crystal-clear waterway teeming with life above and below the surface. We were fortunate enough to paddle our two-man kayak right past a manatee and her calf, as well as seeing vultures, pelicans, osprey and the gangly anhinga, or snakebird as the locals call it. The river is pitted with turquoise spring vents that look incredibly enticing but generally have warning signs not to venture in, as others have tried and sadly not found the exit, due to disorientation.

Approaching a spring vent on the Chassahowitzka River. Photo: Craig Skills

South of Craig and Diane's, we took a wide berth of Tampa, keen to avoid the city traffic. We had arranged to stay with Pam and Tom in North Point. Google Maps proposed two routes. One was about fifty miles and one was eighty miles. We took the fifty-mile option and had our only Google Maps nightmare of the trip.

As we turned off Highway 72 the tarmac disappeared and was replaced by gravel that quickly turned into thick sand, three inches deep. We became sceptical about the route but carried on in the knowledge that this track was only a few miles and the alternative was an extra thirty miles. An hour later we had only progressed two miles. The sand was so thick we were reduced to pushing our heavy loads. We were also growing concerned about the increasing number of warning signs about live shooting in the area, which weren't helped by the odd crack of a shotgun nearby.

Our weariness turned to hope as we could see an open gate ahead and the terrain beyond was hard packed sand. As we approached the gate an old pick-up overtook us and then a mountain of a man wearing only dungarees jumped out and closed the gate. He jumped back into his pick-up, span the vehicle around and set off back towards us. Distraught, but extremely nervous, we flagged him down and asked if we could cycle through. "Nope, private", were the only words he ever spoke to us.

We explained that Google Maps had routed us down there and that our only other option was to push our bikes back through the sand for another hour and then cycle an extra thirty miles to North Point.

Without speaking man-mountain leapt out again, flipped open the tailgate and lifted our fifty-five-kilogram bikes into the back of the pick-up as easily as he'd closed the gate in front of us. We looked at the two-man cabin. It already had three people and two dogs in it. Without saying anything we hopped in the back of the pick-up ourselves and hoped we were going back to Highway 72. After a short, bumpy ride we were back on tarmac.

See you later alligator

Man-mountain nonchalantly lifted our steeds out of the pick-up, we jumped out and he sped off.

Hearts pounding, we spent a few minutes contemplating what had happened, wondering why the man had driven down the road just to close the gate. Had someone seen us and called him? Or was he on his rounds and it was just fortunate timing that he closed the gate moments before we were about to ride onto a live shooting reserve? We never did find out.

We chugged on, cursing Google Maps and thanking the Gods of Cycle-Touring that we'd emerged unscathed thanks to the kindness of the taciturn man mountain. A few hours later we arrived at our hosts in North Point. We told Tom about our plans to head down to Key West. He warned us that it was expensive and said that we may be able to stay with his friends Rupert and Martina. We left Pam and Tom's house, hopeful that their friends might host us and headed to our next stop, Fort Myers.

***

Back in Louisiana I had started experiencing problems with one of the clips on my Arkel front panniers. The bag kept coming off the bike and I had bodged a workaround by securing the pannier to the rack with zip ties. This was fine while riding and had served admirably for over 500 miles, but made taking the pannier off the bike irritating. I had contacted the company directly, and had been blown away by their response. Knowing I couldn't return the bag, the company diagnosed the issue over email and identified the part needed. They asked me our intended route and, using Strava to track our live location, determined that they'd send the part to a shop in Fort Myers called Coastline Cyclery, which happened to be owned by an English lady.

The staff at Coastline were amazing, showing me how to fix the pannier and recommending a great route to our next hosts Beverley and Alastair. Kasia (who we met back in Tobermorey) had put us in touch with Beverley and Alastair, describing them as 'awesome people' that we should definitely stay with. Having already put us in touch with her friends in Penticton, we hadn't doubted Kasia for one moment and gladly accepted an invitation from her friends in Fort Myers.

Beverley and Alastair are snowbirds, spending winter in their Fort Myers condo and summers in their house near Lake Erie. We had a wonderful barbeque that evening, sharing stories and hearing about Alastair's Land's End to John o'Groats bike ride that he undertook fifty years ago. The next day our hosts took us to Sanibel Island for a tour of the Ding Darling National Wildlife Refuge, which allows cars to slowly drive around the four-mile road that cuts through the park. We saw countless spoonbills that, with their pink feathers, resemble portly flamingos. We also saw lots of bizarre tree crabs crawling all over the mangroves that are such a crucial part of the park's ecosystem.

The following day, our hosts took us around San Carlos Bay on their boat and then bought us a slap-up seafood platter at a local restaurant. After our seafood banquet we had a short ride to Collier-Seminole State Park. Beverley had arranged for a local news channel to run a story on our trip and their anchor Channing Frampton met us there the next morning. We watched with amazement as Channing singlehandedly turned our camping pitch into a temporary outdoor studio. He then interviewed us both and filmed us riding for some low speed 'action shots'. It was all very exciting and we set off into Florida's subtropical wetlands, wondering if we'd make it onto the telly.

We spent that evening sleeping at the Big Cypress Oasis Visitor Center. Fletcher, a Warmshowers host and wildland firefighter, was sharing one of the boarding stations with two of the local rangers, Brian and Terrence. Brian taught us much about the local wildlife, with particular reference to the problems created by a recent insurgence of Burmese pythons. The rangers suspected that the snakes were being abandoned by exotic pet owners that hadn't anticipated their enormity:

they can grow to over twenty feet in length. Although classified internationally as a threatened species, the invasive reptiles are thriving in Florida's wetlands, feasting on deer and wading birds that are ordinarily the prey for the Florida panther (the most endangered cat in North America).

Five minutes of fame with Channing Frampton

I was excited when Terrence tottered into the kitchen wearing a Murder City Devils T-shirt. Keen to meet another fan of the band, Terrence proceeded to hog the music all night insisting we all listen to the band's back catalogue while eating, drinking and playing cards.

Fletcher was also hosting another cycle tourer from Quebec called Alice. The next morning Alice joined us for a ride to Shark Valley; a fifteen-mile paved loop off the Tamiami Trail that you can cycle around and see hundreds of alligators. Before we left the ranger station, a couple approached us to ask if we were the cycle tourers from the telly. We hadn't seen the clip yet but the couple assured us we were on the evening news, wished us well and insisted on giving us a voucher for a free meal at a burger chain. Curious about the clip, but unable to see it due to a complete lack of internet signal, we set off with Alice to visit Shark Valley.

Alice somehow wangled us free entry to Shark Valley – maybe playing on our newfound celebrity status – and our visit there was a real highlight of the whole trip. Riding around the loop, in constant close contact with these dinosaur-like beasts, was intimidating at first. However, we quickly realised that, contrary to our preconceptions, they were far more interested in finding the perfect sunny spot and lying still than they were in tearing lumps out of wheeled prey. Amongst the sightings were tiny baby alligators swimming and monster fourteen-foot alligators from the viewing tower. One outrageous alligator did its business right in front of us as we cycled by inches from its steaming deposit.

Alligator scat

Thankfully, Rupert and Martina agreed to host us in Marathon on the Florida Keys. Over the 168 miles we rode from Big Cypress to Marathon, we climbed the grand sum of 153 feet. To put into context how flat that is, there is a hilly road in Bristol called Park Street which climbs 136 feet in half of one mile. Florida is exceptionally flat and that's another reason why Ties loved it.

The night before we arrived at Rupert and Martina's we struggled to find anywhere to stay. The campsites were charging a minimum of $100; which was quadruple what we'd paid for any campground in the US up to that point. We saw a picnic bench in the car park of a medical supply shop and pulled over to assess our options. Then Ties said "I wonder if we can camp here."

With that she confidently strode into the shop and the owners said it was no problem. Furthermore, there was a hose round the back that the owners said we could use for drinking water. Fortune favours the bold.

The ride from the medical supply shop to Marathon was amazing. The archipelago is so narrow that you're almost always riding next to the sea and there are beaches everywhere. Unfortunately, a pleasant beach view is mostly all you get, as despite all beaches in Florida technically being public land, the abundant residential and leisure complexes consume them within their gated developments. We were therefore delighted to come across Anne's Beach (fifteen miles into the day's ride), a tiny cove with a wood deck and public shower. Just what we needed after a night sleeping in a car park.

We found our hosts' home and were immediately greeted like old friends. Rupert and Martina's garden was dominated by a rattan tiki bar and swimming pool. Ties' eyes were on stalks and when Rupert asked if we'd like to stay a few days there wasn't much to consider.

Our hosts had the only licence from Key West to Miami to carry out motorcycle tests. The following day, as Rupert and

Martina examined would be easy riders, we wallowed in their pool and tended to some travel admin. Rupert said we should spend the day relaxing because that evening we'd be going to something called a Superbowl Party.

On 5th February 2017 quarterback Tom Brady, arguably the 'greatest NFL player of all time', was leading his New England Patriots team out against plucky underdogs the Atlanta Falcons. I don't know anything about NFL – I didn't even know who Tom Brady was until that day – and it all sounded more like the plot of the next *Star Wars* instalment rather than a game of football to me. Rupert, a ponytailed motorcycling Austrian, confessed that he didn't know much about the game either but it would be a fun night with lots to eat and drink. He was right. There was more food than I imagine the Patriots and Falcons could even consume, let alone the elderly crowd whose Superbowl Party we were gate-crashing.

The underdogs sped to a 28-3 lead. Brady and his Patriots looked pretty shoddy to me. Rupert wasn't impressed either and confidently proclaimed the Falcons noble winners. Keen to entertain us in his tiki bar, Rupert suggested we spare Brady the humiliation of our watching the end of the game and hightail it back to his. By the time we returned, someone at the party had texted Rupert to say Brady had led his motley crew of footballers to overturn the biggest deficit in Superbowl history. We all agreed that was jolly impressive, shrugged, and retired to our host's tiki bar to discuss tomorrow's proceedings.

Rupert knew how much I wanted to ride the Seven Mile Bridge but he also knew that we were unlikely to find a campground with availability in Key West. He also confirmed that even if we did find a pitch in Key West, it would cost at least $150, being the most sought-after location on the archipelago. Our hosts said we were welcome to stay another night and that they would be happy to spend the following day giving us a tour of Key West in their car. I had already conceded I wasn't going to be able to ride over the fabled Seven Mile Bridge and I was therefore delighted that our hosts would use their local knowledge to show us all the best sites.

The next day, en-route to Key West, we stopped on Seven Mile Bridge to admire Bahia Honda State Park. With its lush, jungle-esque setting right on the water's edge, it was clear to see why the campground there was booked so far in advance. We also saw the derelict original bridge, which runs parallel to the Seven Mile Bridge.

Completed in 1912, the original Knights Key-Pigeon Key-Moser Channel-Pacet Channel Bridge (as the old Seven Mile Bridge was formally called) was the vision of businessman Henry Flagler, who dearly wanted to connect the port at Key West to the US mainland via a trainline. At first, the project was a huge success and facilitated a burgeoning trading partnership between the Caribbean and Florida. However, competition from the shipping industry started to undermine the economic viability of the trainline. Then, in 1935, a catastrophic hurricane pummelled the Florida Keys, with 200mph winds and seventeen-foot waves. Hundreds of lives were lost and significant sections of the bridge destroyed. In 1982 the authorities completed the construction of the current Seven Mile Bridge, and Flaglers bridge was closed for repairs.

Our hosts explained that despite it presenting a safety concern, the old bridge is constructed from such thick concrete that the authorities cannot dismantle it, and it's become a historic landmark. In 2014, plans were announced to spend $77 million restoring Flaglers bridge and it is scheduled to reopen in May 2022.

Down in Key West we enjoyed a few beverages in the bars of Duval Street, visited Ernest Hemingway's house and Truman's Little White House. We discovered that the latter was the de facto US command centre, and allies meeting place, for many of the Cold War conflicts.

Despite our initial enthusiasm the huge queue meant we quickly lost interest in having our photo taken at the 'Southernmost Point of the Continental US'. The tourist trap is essentially an elaborately painted concrete buoy and we opted instead to take a snap at the seemingly unpopular Mile 0 marker for Highway 1.

***

Our next stop after the Florida Keys was Miami. We knew we wouldn't make it in a day from Marathon and were curious where we'd find accommodation in between. On the day we left Rupert and Martina, as we nudged over the fifty-mile mark, we started scouting for campgrounds. We'd conceded it would be an expensive night but even with this willingness to shell out big bucks all the campgrounds were either full or didn't do 'tenting'.

We'd heard stories about cycle tourers staying at churches and fire stations so we started asking if we could stay but with no success. Eventually, as we re-joined the mainland at Manatee Bay, we spotted a dilapidated looking sign for a campground. We weren't sure if it was open or not but as we entered the owner spotted us and called us over, explaining that he'd recently bought the place and was renovating it. The owner said the main facilities weren't open but we could use his shower and toilet if we wanted. This seemed too good to be true and, taking a deep breath, we asked him how much this would set us back. "Oh, seeing as we're not officially open, call it $15."

We thanked him profusely and enjoyed a peaceful night being rocked to sleep on a floating jetty.

The Florida Keys archipelago really was as prohibitively expensive and unsuitable for cycle touring as we'd been warned. However, due to the kindness of strangers and the hospitality of Rupert and Martina, we'd been fortuitous enough to experience its beautiful natural landscape and awe-inspiring manmade wonders. 'Kokomo' might not be a real place but the Beach Boys were right when they told us that the Florida Keys is 'where you wanna go to get away from it all'.

CHAPTER THIRTEEN

OH CAROLINA

8th February – 4th March 2017
Distance ridden this chapter: 1,122 miles / 1,806 kilometres
Total distance ridden: 9,712 miles / 15,630 kilometres

By the time we left Manatee Bay we'd spent three weeks riding 1,000 miles across Florida's Panhandle, down its west coast and around the Keys archipelago. Our final east coast stretch of the Sunshine State started with a visit to my schoolfriend Harriet, who had relocated to Miami.

Harriet had a glorious apartment overlooking Miami beach. She was unfazed by our unwieldy rigs and dishevelled appearance, having hosted her dad several times as he embarked on his own Brompton-powered folding cycle touring adventures around the US. Hours after arriving, Harriet showed us to the cool bars, restaurants and street art of Miami's

Wynwood neighbourhood, and the following morning we set off on a mission to replace my broken saddle and explore as much of Miami's art deco architecture as possible.

I had noticed a split in the top of my saddle that was only going to become bigger. Buoyed by our recent positive experiences with other equipment providers I contacted the saddle manufacturer and they said I needed to find a local dealer. That was easier said than done on the Florida coast, but we suspected we'd find one in Florida's biggest city and fortunately there was a dealer right outside Harriet's apartment block.

When we explained our plight to the shop assistant he started grinning and said "you're in luck, see that guy, he's Florida's representative for the company who made your saddle."

I couldn't believe it. This was the first time the representative had ever been to that store and we happened to turn up asking for a replacement saddle. We were told that if we visited a bike shop ten miles away in the Coral Gables area they would help us out. We rode to the Coral Gables shop and were casually told to help ourselves to whatever replacement saddle we wanted. It was a surprisingly informal process, and another great customer service experience.

Now that my saddle was sorted, we headed for South Beach and a self-guided tour of Miami's famous art deco hotels. The people-watching along the bike paths and busy streets was excellent. South Beach had it all, from the preening bodies at Muscle Beach, to the beautiful people posing on the beach volleyball courts, to the slick supercar drivers that were clinging onto the 1980's Miami Vice style.

North of Miami we followed the A1A coast road as closely as we could, all the way to the border of Georgia. The A1A is a quiet, scenic road, with either a bike path or healthy shoulder that makes it ideal for cycle touring. This route, due to its winding nature and frequent bridge crossings, is substantially slower than more direct routes such as Highway 1. However, the spectacular views of the Atlantic Ocean, and the pleasure of traversing the coastal bridges, made the extra miles worth riding.

Miami bike path

***

Eventually, the A1A deposited us at Cape Canaveral, where we'd arranged to stay with Jocelyn. Jocelyn was no stranger to cycle touring, having spent years riding with her dad across five continents, through thirty-seven countries. It was an emotional time for Jocelyn and her family as her dad had just left Cape Canaveral to return to the South Pole for another stint as a Satellite Communications Engineer on the US Antarctic Programme. Despite this, the family treated us like long lost friends, plying us with local craft beers and treating us to a delicious meal at the local pub.

The bridge that connects Cape Canaveral to Merritt Island is a rare stretch of the A1A, upon which bikes are not welcome. We hadn't known this beforehand, but fortunately Jocelyn's brother was better informed, and insisted on driving us to the other side. Our goal for the day was to visit NASA's Kennedy Space Center. We had no idea the iconic complex was located in a national wildlife refuge and as we rode across the barrier island, we passed numerous alligators and raccoons.

Ever since our first rest day of the trip – in Tobermorey, Ontario – we'd discovered that tourist attractions are unfeasible on a bike tour. This wasn't just due to the state of the Pound after Brexit – although obviously that didn't help – but also a practical issue with bike security and timing. We'd been incredibly fortuitous on many occasions with hosts offering to show us the sights and drive us to landmarks throughout the trip, but visiting attractions whilst in riding mode was tricky.

This was true of the Kennedy Space Center. I'm not sure what we thought we'd do when we arrived but basically we rode up to the sign near the entrance booths, waited for the hordes of other visitors to have their photos taken and then did the same ourselves. An initially helpful employee in matching baseball cap and bumbag let us know that they had a secure storage for our bikes and we briefly contemplated visiting. That is until we discovered it would cost us $120. We politely declined and he equally politely suggested that we remove our bikes and ourselves from the premises!

***

A few days later we rolled into New Smyrna Beach, where we'd arranged to stay with Lynn and Hank, parents of our friend Greg in Toronto. We had a relaxing rest day hanging out at the pool and walking on the beach. Lynn and Hank treated us to a fantastic stay and took us out for a scrumptious seafood feast.

As we passed through Daytona Beach, we thought we were going to see Trump himself. Florida was readying itself for Trump's first post-inauguration rally. Although the event was due to be held the following day 100 miles south, in Melbourne, the welcome committee had done an admirable job of building excitement for Trump's arrival. Daytona Beach was awash with billboards hailing the recent induction of the forty-fifth US President and welcoming him to the state that emphatically embraced the MAGA movement.

We spent that night in St Augustine Beach and the following morning paid an early visit to Castillo de San Marcos, a seventeenth century fort and one of the oldest stone buildings in the US. The stone fort was built by the Spanish to bolster wooden defences – originally installed in response to attacks from infamous Devonian Sir Francis Drake, in the sixteenth century. In the early eighteenth century the Brits twice attacked the fort, eventually occupying it in 1763 as a provision of the Treaty of Paris. Spain regained control in the early nineteenth century and then the Americans took custody in 1837. The fort was traded back and forth between the Union and the Confederates during and after the American Civil War, and was eventually demilitarised in 1933.

After our culture trip in St Augustine we continued north and discovered that we'd badly timed our penultimate night in Florida, owing to a busy Saturday of fully booked campgrounds. We decided to take the ferry to Fort George Island and found a friendly Circle K gas station that let us camp at the side of the building. While not overly glamorous, this wouldn't be the last gas station we stayed at, as we discovered their benefits, including twenty-four-hour toilets, drinking water and Wi-Fi.

On our last day in Florida, we left the A1A to ride the boardwalks and trails through the fertile Timucuan nature reserve. The tranquillity of Timucuan lasted until ten miles south of the border with Georgia where the leafy bike paths ran out and we reluctantly joined the dreaded Highway 17, which would be our main route for over 300 miles.

The reason it's so terrible is, unlike most roads in the US, there is a narrow shoulder and the shoulder is taken up with inverted rumble strips to wake up weary drivers who veer off the highway. This gives cyclists two options: endure the constant juddering of the rumble strips in return for the safety of the shoulder, or take your chances with the traffic on a major highway. We chose the latter, which was hair-raising but ultimately, we survived and it was far more comfortable than the alternative.

Our first night in Georgia was spent at a dilapidated but extraordinarily inexpensive motel in Woodbine. The tired establishment was the cheapest motel in the US of our trip, but the expression 'you get what you pay for' has never described somewhere so accurately. The water was brown, not a single power socket worked and the stench of cigarettes put us off our porridge. The place had seen better days.

The next day we made it just past the small community of Eulonia and spent another night sleeping next to a gas station. The owner was friendly and even gave us the Wi-Fi password, in exchange for purchasing cold beer and Chips Ahoy cookies. We went to sleep, aided by Yuengling lager and the knowledge that we'd be staying with a friendly host near Savannah the following day, the 21st February 2017. Yuengling is an uncommon tipple in the UK but in the US it's hugely popular, partly due to the fact that it's the biggest American-owned brewery in the US.

***

Savannah played a key role in our trip. To explain why I need to rewind a few days. As we were relaxing in our motel at St Augustine Beach I had received a message from our friend Jon, who we stayed with in Renton back in October. Homeland Security had sent a letter to Jon explaining that our visa extension requests had been reviewed – twenty-seven days after our visas had expired – and they had many questions and requests for evidence. There was also a tricky logistical restriction attached to the request for more information; we had to send back an original copy of the first page of the letter Homeland Security sent to our friend Jon.

Jon had the genius idea of sending the letter to a Post Office on our route for us to collect. We just had to try to guess which town would be best to pick it up from, given our average daily distance (fifty miles) and the time we thought it would take for the letter to arrive. We also had to factor in the deadline – 12th March 2017 – Homeland Security had attached to the request for information. Eventually we settled on Savannah and were delighted when a Warmshowers host called Austin agreed to host us for two nights, so we could collect the letter on our rest day.

After another day on the appalling Highway 17 we arrived at Savannah, to be greeted by a smiling Austin on his driveway with two ice-cold craft beers that he handed straight to us. Austin prepared pizzas from scratch and laid out a selection of home-smoked delicacies for us to enjoy. Our host was working the next day but insisted we borrow his car so we could travel to the Post Office and explore central Savannah.

Like Tucson in Arizona, we knew nothing of Savannah, but it would prove to be one of our favourite cities of the trip. Our first stop was the Post Office, where we picked up the letters Jon had forwarded from Homeland Security. We then spent hours wandering the streets, drinking in the culture and colonial history of laid-back Savannah. After a delicious Southern barbeque lunch, we continued exploring until the heavens opened and dumped a biblical rainstorm on us. In minutes we were completely drenched. So wet, in fact, that my iPhone, which was in a waterproof pocket that had fended off a month of rain on the Pacific Coast, stopped working completely.

I wasn't initially too bothered about the phone – reasoning that I could salvage it by drying it out in a bag of rice – but when we returned to Austin's car it dawned on us: how would we find our way back to Austin's? Ties didn't have data on her phone and we had blindly followed Google Maps to navigate to central Savannah. We decided to try to wing it. After an hour of driving in circles we had it upon a new technique; find a McDonalds, blag Wi-Fi in the car park and take screenshots of the route. This sort of worked and we eventually found our way to Austin's by making frequent stops at fast food chains to pilfer Wi-Fi until we recognised Austin's neighbourhood.

That evening Austin had plans and he gave us strict instructions to help ourselves to his extraordinary craft beer collection while he was away. He also said we could use his computer to create a response to Homeland Security's request for evidence. The latest information requested was impossible to provide and I decided to take a different approach to my previous pandering to Homeland Security's demands. I wrote a brusque letter explaining that I couldn't provide both proof that I lived abroad and proof of my US residence, as this was both contradictory and erroneous. I also explained that I couldn't provide the original copies of supporting evidence to any of Homeland Security's requests as I had already sent them to Homeland Security in response to one of their many previous letters.

I realise this sounds truculent, but I was fed up with the process and not in a conciliatory mood. I had lost all patience and wanted certainty about how the remainder of the trip would pan out, be that in the US or not. Ties was upset at the likely prospect that Homeland Security, under Trump's stewardship, would likely view my bolshie tone unfavourably. However, she understood my frustration and concurred it was impossible to meet their request for evidence. We decided to stick to my terse reply, telling Homeland Security I couldn't provide a single piece of information they requested, and let the outcome of this bold new approach determine how we completed the trip.

***

Unfortunately the patented 'bag of rice' technique had failed to revive my iPhone, and so the following day we headed to Charleston to explore the city and find a replacement. There weren't many accommodation options between Savannah and Charleston and we ended up spending two consecutive nights camped outside fire stations. Using Ties' phone, we managed to sponge enough Wi-Fi here and there to arrange to stay with a Warmshowers host in Charleston called Lenny.

Lenny explained that he was on a trip in the mountains and would return after our estimated arrival time. Despite never meeting us, Lenny gave us instructions of how to enter his house and told us to make ourselves at home until he returned. This was the first time someone had given us entry to their house unaccompanied, and we were humbled and dumbfounded by the trust of complete strangers.

After picking up a replacement iPhone, we had a great time exploring Charleston's historic market, meandering around the city streets and relaxing at the waterfront. Lenny took us out for a tasty meal in the evening and made me gooey-eyed by showing me his mint Merlin titanium bike. Back in the early 1990's, my friend Neil and I would have sleepovers and stay out of mischief by geekily specifying dream bikes. In those pre-internet days, magazines like *Mountain Biking UK (MBUK)* would carry adverts for bike shops that would be six pages long and painstakingly list all the components, prices and weights. My dream bike set up would invariably have a Merlin frame but I hadn't seen one in the flesh until our stay with Lenny. Nowadays, Neil, and my aforementioned friend James (he whose book introduced me to the Pacific Coast route), have their own brand of dream bikes (Tora Cycles) that appeared in the October 2021 edition of *MBUK*.

As well as being a connoisseur of fine bicycles, Lenny was also a long-term cycling campaigner. His efforts in the 1990s helped secure funding for the two-mile bridge that guided us out of Charleston. The Arthur Ravenel Jr. Bridge, as it's formally called, is a spectacular feat of engineering with an interesting history. In 1995, the authorities assessed the safety of the Grace Bridge, which had provided onward transit from Charleston over the Cooper River since 1929, as four out of one hundred. It doesn't take a safety expert to realise that four percent is sub-optimal. US Congressman Arthur Ravenel Jr. was enraged and ran for election in 1996 with the sole objective of replacing the decrepit structure with a new bridge. He won and, with the additional efforts of local campaigners like Lenny, the new bridge was opened in 2005. Locals celebrated by taking to the old bridge one last time in the form of a 'Burn the Bridges' run and a parade of 1929-era cars.

Not long after our ride over the much-celebrated bridge Ties' bike let out an almighty bang, followed by the heart sinking 'pffffff' of deflating air. She told me that as well as having a puncture she couldn't move her bike at all. "That's because one of your spokes is sticking through your tyre" I diagnosed, unhelpfully.

After some initial shock at what had happened it dawned on us that this was our first spoke failure in 9,400 miles.

Spokes are heroic bicycle parts. They weigh six grams. The same as a biro. They use the hub – the middle bit of a bicycle wheel – as an anchor to brace the rim. If the spokes aren't equally tensioned, they can't spread the weight of the rider equally and the wheel either buckles or the spokes snap. In Ties' case the nipple – the tiny bit of metal that holds the spoke to the rim – had failed and the spoke shot through the rim tape, tube and tyre.

As previously mentioned, I'm not the world's best bicycle mechanic and my only support was to unlace the spoke from the wheel. This left Ties with a Pringle-shaped wheel and a challenging ride. Despite the constant rubbing of her rear tyre, on the bike's rear triangle, Ties gallantly rode on for fifty miles, before we found a bike shop that could true the wheel.

The remainder of our time on the mainland of South and North Carolina was dominated by the heinous Highway 17. The stretch from Charleston to Jacksonville was also the last part of the trip were the hourly dog-induced spinning classes took place. As we contemplated our onward route we were sad to leave behind the famously warm hospitality of the Deep South, but were also grateful that we'd be saying goodbye to all the charging canines.

CHAPTER FOURTEEN

GEORGE WASHINGTON'S A MACKEM?

5th – 23rd March 2017
Distance ridden this chapter: 644 miles/ 1,036 kilometres
Total distance ridden: 10,356 miles / 16,666 kilometres

The boneshaking inverted rumble strips made our Highway 17 route through Georgia and the Carolinas tedious, but straightforward. There wasn't a whole lot of route planning to consider and we navigated the 250 miles from Charleston to Jacksonville with ease. Our onward route from Jacksonville to Boston however was a different matter. There were so many options for navigating the islands and bridges around the Delmarva Peninsula, as well as important choices about whether to visit Washington DC, how on Earth we'd tackle NYC and what route we'd take through New England.

Our first route-based decision of the Atlantic Coast came at

Jacksonville, North Carolina, where we could continue inland on Highway 17, or take the slow and scenic option via the Outer Banks (OBX). The OBX is a string of barrier islands off the coast of North Carolina and Virginia. Since 2020 the OBX has become well known thanks to the teen TV series of the same name. In 2017 we hadn't heard of the OBX at all but fellow travellers had waxed lyrical about the remote picturesque islands, inhabited by semi-wild horses and important historical landmarks. Given that the alternative choice was our old nemesis Highway 17, it wasn't too hard a decision to make.

Riding the OBX was exciting but required careful planning. Fortuitously, our last night before heading to the OBX was spent with hosts Tom and Nancy, at the small town of Cape Carteret, who were a treasure trove of local knowledge. Over an appetising homecooked dinner, Tom and Nancy gave us all the advice we needed about how to navigate the islands and their infrequent ferries.

We had an extra early start the next morning and then raced the sixty miles to Cedar Island, to catch the last ferry of the day to Ocracoke Island, which left at 4:30pm. Ocracoke is small – only fourteen miles from one end to the other – but it is delightfully undeveloped, with a rustic, wild charm that belies its diminutive size. From North Ocracoke we hopped to Hatteras Island where we visited the Cape Hatteras Lighthouse, the tallest brick lighthouse in the US. With its black and white humbug-like paintjob, the lighthouse is a striking landmark with a fascinating past.

Originally constructed from sandstone in 1803, the lighthouse came under attack from Confederate forces in 1861 and since then has suffered from the erosion of the coastline and a battering from the intense OBX winds. In 1870 US Congress approved funding for a replacement brick lighthouse and the original structure was abandoned. By 1933, further erosion left the new lighthouse only 100 feet from the shoreline; down from 1,500 feet in 1893. In 1935 the second lighthouse was abandoned and was replaced by a third steel skeleton structure in a different location. Over the next fifteen years the Civilian Conservation Corps created manmade sand dunes, enabling the beacon to transfer back to the second brick lighthouse in 1950.

By 1980 further erosion left the brick lighthouse precipitously close to the shoreline. In 1999 the entire lighthouse was lifted off the ground and moved 2,900 feet to a safe location. Standing at the bottom of the 198 feet lighthouse, it was astonishing to imagine how they achieved this audacious feat. The actual move took twenty-three days, cost $12m and was facilitated by using hydraulic jacks to diligently inch it along a specially made track.

After marveling at Cape Hatteras Lighthouse, we rode over to Pea Island, a national wildlife refuge that has itself fallen victim to the wild OBX weather. Pea Island hasn't always been a true island, with the shifting sands opening and closing the inlet that separates it from the adjacent Hatteras Island. From 1945 to 2011, Pea Island was actually the northern point of Hatteras Island, making it a land-tied island. When *Hurricane Irene* struck in 2011 it caused huge damage to the OBX and reopened the inlet, making Pea Island a true island once more.

From Pea Island we rode over the huge bridge that spans Oregon Inlet to Bodie Island. We had arranged to stay with Warmshowers host Pat in Kill Devil Hills, made famous as the site of the Wright Brothers inaugural flight of their Wright Flyer aircraft in 1903. Kill Devil Hills was chosen for its persistent strong winds, remoteness and abundance of sand dunes, which the Wright Brothers figured would be ideal for a soft crash landing. While the winds and sand dunes remain 120 years after the Wright Brothers first attempted flights, the area is now far less remote with swathes of houses and resorts. We stopped off at the memorial, which has pieces of the Wright Flyer that were taken to the moon by Buzz Aldrin and Neil Armstrong in 1969, commemorated as the 'First Manned Lunar Landing: Kittyhawk to Tranquillity Base'.

Wright Brothers Memorial at Kill Devil Hills, inspiration for Bruce Dickinson's song of the same name

Pat's house was an architectural marvel, full of interesting features including a 'microgreens tower garden', dedicated dog bath and pot filler over the hob. When we rang the doorbell, it called Pat's phone and she was able to see it was us and then let us in remotely. The back of the house had a huge veranda that overlooked the tranquil Kittyhawk Bay. Pat offered to let us stay an additional night and we spent the following day hanging out on the beach with her adorable Miniature Pinschers and black Jack Russell Chihuahua cross. As we ran around on the beach in glorious twenty degrees Celsius sunshine, we could hardly believe that snow was forecast the next day.

***

Our rest day at Kill Devil Hills gave us time to contemplate our onward route. We'd become sold on the Delmarva Peninsula, so named for the three states it encompasses: Delaware, Maryland and Virginia. However, we were also intrigued about visiting Williamsburg, Richmond and

Washington DC and worked out that we could do both by joining the Delmarva further north.

Williamsburg is a small city, known for its colonial history and critical battlefield site during the American Revolutionary War. We stayed with Warmshowers host Nick who took us to the Yorktown Battlefield, where General George Washington won independence from the Brits in 1781. It was a sobering experience, made a trifle awkward when the tour guide asked our group where we were all from.

Later that evening Nick plied us with local craft beers and told us all about his role as an animal rescue pilot for the charity Pilots N' Paws. In his spare time Nick would fly around the country rescuing animals that would otherwise be euthanised. This included one occasion where Nick managed to single-handedly transport thirteen dogs (including an Irish Wolfhound) and four kittens in his tiny two-seater *Cherokee* plane. Another story Nick told us involved him flying down to an airstrip in South Carolina to collect sixteen dogs that his friend had rescued from Kentucky. The dogs were bundled into Nick's pintsize *Cherokee* and flown straight to a loving foster home in New Jersey. The world needs more people like Nick.

***

The weather on the east coast had been mercifully forgiving. It was a bit chilly, but it was mid-March, so we were still transitioning to Spring. Despite this, the weather forecast on the day we left Williamsburg was concerning: it was predicting a Nor'easter, an East Coast storm with north-easterly winds and 'ice pellets' (the US term for sleet). That sounded cold.

The storm was forecast to dump heavy rain and snow on a huge stretch of the East Coast, all the way from Virginia to Quebec. So many blizzard warnings were issued by the US National Weather Service that the storm was dubbed the *Blizzard of 2017*.

We left Williamsburg with our warm clothes at hand and set

off for the Virginia Capital Trail, a bike path that joins the old colonial capital with Richmond, the modern-day capital of Virginia. For over fifty miles we rode along dedicated paved bike paths that wound through ancient forest, over verdant wetlands, passed historic plantations and more American Revolutionary War battlegrounds. It's a truly breath-taking trail and – even as the temperatures plummeted close to freezing – was my favourite US bike path of the trip.

The remarkable Virginia Capital Trail

Richmond looked like a fun place. As we searched for our host Robbie's bike shop, we admired the welcoming bars and restaurants, and again wished we had the budget to explore. We found Robbie's shop and rode in convoy to his house. Robbie confirmed that the Nor'easter was starting in earnest that evening and offered us an extra night at his house. We gladly accepted but had no idea what to do with our bonus day indoors. Robbie recommended the sci-fi coming-of-age TV series *Stranger Things* and we spent the next day binging the whole of season one. When Robbie came home from work that day he found us square-eyed, glued to the telly and seemed rather pleased with himself that we enjoyed it so much.

Although the *Blizzard of 2017* wasn't as extreme as forecast, the Governor of Virginia declared a state of emergency, as did his counterparts in New York, New Jersey and Maryland. Virginia lined up 4,500 snow ploughs and New York City shut the subway. Trump postponed his first meeting with Germany's Chancellor Angela Merkel, where the two leaders had planned to discuss solutions to Russia's land grab of the Ukraine. The political summit was so highly anticipated that when it was cancelled Merkel travelled to Berlin airport to reluctantly tell reporters on their way to Washington DC that "The trip is cancelled. That is not a joke."

Our trip would not be cancelled. Not only had we been hardened by much graver weather than this naughty Nor'easter, we'd also received some wonderful news from our friend Jon in Renton. Our visa extensions had been approved! Two months after they'd originally expired, and after I'd written to Homeland Security saying they couldn't have any of the information they requested. A triumph for good old-fashioned British belligerence. Our visa ordeal was over, we were feeling invincible, and we powered on northbound towards Washington DC.

Snow was piled everywhere. Those Virginia ploughs had clearly done their job and made life plain sailing for us. There was no way we were going to camp in this weather and we were fortunate to be hosted by some great characters en route to the US capital.

Our first stop was in Doswell where we stayed with a Quaker called Dave. His wife Kathleen was holidaying in Mexico, but despite this took the time to coordinate between us (she held the keys to the Warmshowers account). Kathleen and Dave had a stunning house in the woods, completely enveloped by trees and bereft of any manmade noise or light whatsoever. Dave cooked us a scrumptious vegetarian dinner and told us fascinating stories about his life as a Quaker.

My ignorance was exposed when Dave said that being a Quaker didn't necessarily make you religious. He was part of a group that observed the traditional Quaker ideals of peace, simplicity, equality and social justice, but did this through a sense of community, as opposed to a theological belief. This was a revelation to us. And seemed to make so much sense. It also appeared, on at least a surface level, to make complete sense: the idea of a community based around common tangible beliefs, yet not anchored by an invisible being or ancient doctrine, appealed to us.

Dave's claim to be the 'best pancake cook on the East Coast' was confirmed the next morning. He blew us away with a delicious carb-laden breakfast with lashings of maple syrup. We wrapped up in our winter woollies and headed back into the lingering chill left by the Nor'easter. Our goal that day was Fredericksburg, where we'd arranged to stay with hosts Bruce and Vikki.

Bruce is a dedicated Scoutmaster. One of those guys who you know has found his vocation in life and is at his absolute happiest when wearing his woggle. In 2013, as I started planning my bike trip around North America, Bruce was also planning the inaugural Eagle Scouts Cycling Across America ride. The following year he led a peloton of scouts who rode 3,800 miles from San Francisco to Virginia Beach. He followed this up in 2018, leading twenty-five scouts on a 4,200 mile ride from Seattle to Washington DC. If you think the logistics of our trip were daunting, take a moment to consider the planning that Bruce had to do. His scouts consumed a whopping 2,000 kilograms of food, replaced 230 inner tubes and rode through four time zones, all in sixty-seven days. Not one to rest on his laurels, Bruce's goal for 2022 is to cycle a perimeter of the US in six months, following a similar route to us. He's aiming to raise $50,000 for the Multiple Sclerosis Foundation, supporting the charity that is improving the lives of those, like Bruce's brother, who suffer from the condition.

It was inspiring to meet Bruce and Vikki and trade Scouting stories from one side of the pond to the other. In addition to the unwavering commitment to Scouting, Bruce and Vikki have also somehow found time to host over 100 cycle tourers. We were taken in despite Bruce's intense work schedule, and us changing dates because of the Nor'easter.

***

On a bike, you can enter and leave Washington DC without touching a single road. We rode into the capital on the excellent Washington and Old Dominion Trail. This surprised us, as we'd heard horror stories about riding in the capital. What surprised us even more was the lack of security or restrictions at any of the high-profile landmarks. It was incredible. First off, we headed straight to the Lincoln Memorial to snap a cheesy photo with the Reflecting Pool and Washington Memorial behind us. Tick. Then we rode right up to the Martin Luther King Jr Memorial. Another photo ticked off. Finally, we rode over to the Washington Memorial, found a bench and brewed a coffee on our highly explosive propane/butane powered stove. No one blinked an eyelid. Fantastic.

We were fortunate with our accommodation in the capital. Pat (from OBX) had told us that her daughter Lisa lived in Washington DC and had moved out of her flat, but still had the keys. It was empty, but all utilities worked and we were welcome to stay for as long as we needed. Lisa's flat was in a fantastic location in the Columbia Heights area of the city and we took the opportunity to have two rest days.

Washington DC's Smithsonian museums are famously (mostly) free, which makes it the perfect rest day spot for cycle tourers. We explored the Museum of American Art, National Museum of African Art, National Museum of Natural History, Hirshhorn Museum and Sculpture Garden, National Air and Space Museum and National Portrait Gallery. We roamed around the gardens and parks, and amused ourselves at how small the White House is in real life. We were able to meet up with Pat one night and enjoy a grand night out, wondering which Congressperson or Senator the people around us were working for. It was an excellent pit stop and yet another American city that we have fond memories of.

Reflecting Pool and Washington Memorial

The US capital has some interesting traits. For one, it is the only place in the country not geographically located in one of the fifty states. It is located in a sort of neutral depoliticised zone called the District of Columbia. Even more intriguing, it has fifteen sister cities around the world. All but one are capital cities, such as Ankara, Bangkok, Beijing and Rome. The one exception is North-East England's Sunderland: yes, *that* Sunderland! Land of the *Mackem* and hitherto self-proclaimed 'greatest shipbuilding port in the world'.

The link between the two places isn't obvious but arises from George Washington's familial heritage in the town of Washington, on the outskirts of Sunderland. The Founding Fathers' ancestors first arrived in Washington in 1183, and five generations of his descendants lived in Washington Old Hall. In 1656, John Washington entered the tobacco trade, invested in a merchant vessel called *Sea Horse,* and set sail for the Colony of Virginia. John decided not to return to England – partly due to the state of the country following the nine-year-long Civil War – and settled in Westmoreland County, in what is now the State of Virginia. Seventy-six years later, John's great-grandson George was born, and would go on to become the first president of the US.

The Washington family crest is emblazoned with red and white stripes, like Sunderland A.F.C's football shirts. There's a theory that the stripey bit of the Stars and Stripes flag has its origins in Sunderland. I am a Newcastle United supporter. Newcastle United are arch rivals of Sunderland A.F.C and I'm not having of this nonsense about those *Mackems* influencing the US flag.

***

We left the capital via the wonderful Anacostia Riverwalk Trail that snakes along the river of the same name, through parks and neighbourhoods that seem like a different world to the glitzy houses, monuments and Smithsonian museums of central DC. We had a short ride that day, an easy thirty-six miles to stay with our host Cindy in Annapolis, Maryland. Cindy is not a cyclist, but her brother Glenn was an avid cyclist and toured extensively. Sadly, Glenn lost his battle with cancer, and Cindy saw Warmshowers as an opportunity to give back on his behalf.

Cindy had a wonderful house with attentive dogs that provided entertainment all night long. When Cindy said dinner was ready, we thought we were attending a dinner party with half the street. She had rustled up a feast fit for all the Eagle Scouts of America, with dishes that the *Great British Menu* judges would be fawning over. We had a lovely evening, chatting with Cindy and gaining insight into the logistics of riding on the Delmarva.

Chesapeake Bay is the largest estuary in the US and separates Annapolis from the Delmarva Peninsula. The four-mile-long Chesapeake Bay Bridge stands 200 feet above the estuary and regularly tops polls as the scariest bridge in the US. Cycling is not permitted, and an enterprising local company has responded by offering cyclists and scared drivers transit over the bridge. When we were picked up, there was a lady in the minibus who was so scared of even being driven over the bridge that she blindfolded herself for the crossing; a journey she makes every day. One of the taxi drivers drives her car over the bridge, she hops in the minibus with disgruntled cyclists and pedestrians, and is reunited with her car on the other side.

Once again on terra firma, we rode through quaint farmlands and flat country lanes into our seventeenth state, Delaware. We spent our only night in Delaware sleeping in a log cabin on the farm of hosts Stuart and Delores. The farm was home to a friendly donkey, goats, chickens, and was well guarded by Sixty, our hosts' enormous Basset Hound. The cabin was exactly what you'd expect your quintessential pine forest hideaway to be like. There was a log burner, a kitchenette, and plush handstitched blankets draped over the bed and sofa. Perfect for the bracing minus-ten degrees Celsius weather. Stuart and Delores made us a wholesome sausage bake the next morning, and we headed east well-fuelled on our breakfast of champions.

We had two goals for our second and final day on the Delmarva: to meet up with our friend Alex and to board the ferry to Cape May, New Jersey. We had first made contact with Alex via Instagram, as he was riding the Southern Tier at the same time as us, albeit in the opposite direction. We hoped to meet at Christmastime, but Alex's route saw him go slightly south of us, via Houston. As we passed through the Delmarva in March 2017 Alex was working as a lifeguard in Ocean City, Maryland. We arranged to meet Alex in Lewes, one of the oldest Dutch settlements in the US. We knew this before we arrived but still couldn't believe the number of Dutch references. The Zwaanendael Museum itself is modelled after the town hall in Hoorn, a city and former Dutch East India Company base, thirty miles north of Amsterdam. The De Vries Monument in Zwaanendael Park commemorates the first permanent European presence on the Delaware Bay. And one of the first in the US.

Alex turned up on his skinny-tyred road bike, making me yearn for a faster steed. We enjoyed a coffee and pastry, exchanged cycle touring war stories and followed his lead to the ferry. Two hours later, we were being warmly welcomed by our host Mark and his family in Cape May, New Jersey. Mark provides cycling tours of the local area and helped us with invaluable onward route guidance. The whole family seemed to be musical geniuses and the house had a laid-back vibe.

This wasn't the image of New Jersey that we'd expected. Cape May was a sleepy beach resort with resplendent Victorian villas and pavilions, and unexpectedly quiet beaches. Mark explained we were well out of season and it was a good time to be in the area. We cracked open another beer, sank into the vast sofa, and fell into a trance listening to Mark's son Christian pluck his acoustic guitar.

These were the experiences that made the trip so special. Experiences I almost certainly wouldn't have had if I'd undertaken the trip alone. I would have doggedly ploughed through the miles, congratulating myself for putting in century-days in the saddle, bunking in bus shelters and missing out on encounters with friendly strangers. Ties completely changed the way the trip unfolded. I went to bed that night content, reflecting on my life-changing holiday to Mexico and dreaming about what unexpected encounters we may have on the famous Jersey Shore.

CHAPTER FIFTEEN

LIVE FREE OR DIE

24th March – 11th April 2017
Distance ridden this chapter: 629 miles / 1,012 kilometres
Total distance ridden: 10,985 miles / 17,679 kilometres

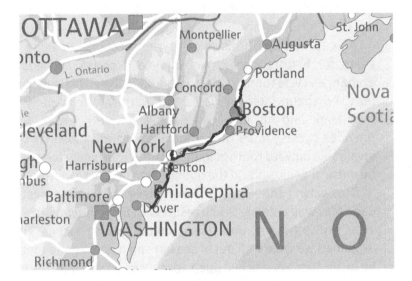

New Jersey was not what we had expected. Back in the UK, our cultural reference points for the Garden State – including mostly limited to *Boardwalk Empire*, *Sopranos* and *Jersey Shore* – evoked images of organised crime, decadence and unnaturally orange people. What we discovered was a quiet Victorian seafront, immaculate boardwalks, woodland bike paths, and miles of wildlife wetlands. It was a pleasant surprise and left us with a good impression of the most densely populated of all fifty US states.

At the small community of Forked River, we spent the night with Diane and Steve, parents of Brent, who we'd first met in

Gualala. They lived in a neighbourhood where each house had its own jetty in the back garden, providing direct boating access to Barnegat Bay, a forty-two-mile stretch of the Atlantic Intracoastal Waterway protected from the ocean by a barrier island. We had a fun evening with Diane and Steve, hearing stories of Brent's transition from child genius to backwoods camping aficionada. We promised Diane and Steve that we'd say hi to Brent in Maine – where we'd arranged a reunion – and then headed north.

We had two options for accessing NYC: the George Washington Bridge or the Highlands Ferry. Although we have a penchant for riding over bridges, the George Washington Bridge is sixty miles north of the Highlands Ferry and we therefore settled on a waterborne arrival into Manhattan's Battery Park. This was a great choice, and it was an expensive but stress-free arrival into the Big Apple.

As soon as we rolled off the ferry, Ties was 'doing an Eddie Murphy' on every corner. There is nothing that can prepare you for NYC. Sure, London is busy, but NYC is a gridded rat run of manic, honking, yellow taxis and all manner of other road-going vehicles. Most roads are lined with soaring skyscrapers that would dwarf The Shard and the sense of urgency makes London seem like *Downton Abbey*.

Aside from cycle couriers – who are as mad as the taxi drivers – people don't tend to ride around Manhattan. For one it's not really necessary, it's walkable, and the public transport is okay, if a bit shabby. But the main reason is the taxis. Zebra crossings and Pelican crossings – or crosswalks and traffic signals as Americans call them – should be used with caution in NYC: more of a guide to when traffic *may* stop, as opposed to a reliable indication that it's safe to cross. The same contempt for traffic laws is levelled at cyclists, who appear to be invisible to drivers of motor vehicles.

Armed with this knowledge, from my previous trips to NYC as a tourist, we chose to give it a go anyway. After a quick stop at the Charging Bull statue – to snap a cheeky photo of Ties tickling its balls – we joined the Hudson River Greenway. Supposedly the busiest bike path in the US, the Hudson River Greenway is thirteen miles of traffic free trails from Battery Park to just north of Washington Heights, where we were staying. If we had wanted to, we could have taken the Hudson River Greenway all the way from the ferry to our host Nick, who was one of the cycle tourers we met in the desert in Texas. This seemed straightforward, but we wanted to ride in the throng of NYC, especially around Central Park.

We left the bike path momentarily to visit the 9/11 Memorial. The footprint of each of the fallen Twin Towers have been replaced with expansive manmade waterfalls – the largest in North America – that give the impression of eternal water cascading into the ground. The serene infinity pools represent the absence of those lost during the horrific events that day: 'although water flows into the voids, they can never be filled'. The monument is hugely thought-provoking and you cannot help feel for the families of victims when you read their names forever engraved around the edge: a poignant tribute, and an experience that neither of us will ever forget.

We re-joined the Hudson River Greenway until Greenwich Village and then decided to brave a ride into the melee of yellow taxis. After a few blocks we joined 6th Avenue – the Avenue of the Americas – and held our breath. Our cumbersome rigs weren't exactly ideal for navigating the traffic, but we made it to Central Park unscathed, having been assisted by a cycle courier who had noticed we obviously didn't do this every day. He escorted us along the busiest stretches, using a combination of hand gestures (some rude), nods and belligerent posturing of his bicycle towards the frenetic taxi drivers encroaching our space.

After the mania of 6th Avenue, Central Park was positively blissful. We rode around the trails with ease, following the lycra-clad roadies on their training laps and a surprisingly high number of horse-drawn carriageways. Leaving Central Park, we whizzed through Harlem without incident and found Nick's apartment in Washington Heights, delighted to see it was next door to both a supermarket and classic American Diner.

One World Trade Center

Nick had three flatmates, and in the three nights we spent in his apartment, we didn't see Nick or any of his flatmates once. Busy people, New Yorkers! To maximise our sightseeing opportunities, we used the extensive bus network to move between famous landmarks, including the Chrysler Building, Grand Central Station, and Empire State Building. We met up with Ties' Uncle Mo and took the free Staten Island Ferry to walk around the Esplanade and visit the Staten Island 9/11 Memorial. In the evenings Uncle Mo treated us to some mouth-watering Persian and Libyan food at his favourite hidden restaurants. He told us stories about where he lived in Astoria, Brooklyn, saying that he had a decent place but recommended that we don't go to the neighbourhood.

***

It is commonly known that New York was originally called New Amsterdam, having been sequestered by the Dutch in 1624. Forty years later, the Brits took over and renamed the city New York, in honour of 'The Grand Old Duke of York', who led the appropriation. A lesser-known fact is that Brooklyn is named after Breukelen, a wealthy merchant town near Utrecht in the Netherlands. The Brits clearly didn't mind the moniker too much and simply anglicised the name to make it more palatable to English speakers.

Before I met Ties, I'd stayed in Brooklyn a few times and really enjoyed the Borough. I was therefore delighted that two Warmshowers hosts – Chris in Greenpoint and Kees in Prospect Park South – had agreed to host us. To get there, we decided to stick to the Hudson River Greenway all the way to Battery Park and then take the Brooklyn Bridge over the river. As soon as we saw the bridge, however, we immediately changed our minds; it was like a human zoo. We watched several cyclists' punitive attempts to cross the bridge, honking and yelling at the dense mass of humanity crowding the 'bike only lanes'. We decided there was no chance we'd manoeuvre our extra wide loads through the crowds. Williamsburg Bridge, two miles to the east, was our saviour. It was empty and provided easy transit to Brooklyn.

We spent three days exploring Brooklyn's diverse neighbourhoods on foot and by bike. We visited the vast Brooklyn Museum – the second largest museum in NYC – perusing its varied art collections, including indigenous artefacts, classical portraits and Judy Chicago's controversial twentieth century feminist exhibit. The latter included a seminal piece called Dinner Party; a giant triangular dining table, upon which guests would be greeted with crockery displaying moulded vulvas. We rode down to Coney Island to admire the rickety amusement park, home of the country's first rollercoaster, and the giant countdown clock that displays the time remaining until the next International Hot Dog Eating Contest.

After a week in the Big Apple we saddled up and headed east. We discovered that there is a ferry from the tip of Long Island to New London, Connecticut, and decided this would provide the easiest escape from New York. All went to plan, except our arrival in New England was greeted with some very Old England weather. It wasn't West Coast standard torrents of rain, but dense mizzle (English-style misty drizzle) that seemed to hang in the air and drench us all the same.

Compared to the rest of the US, New England states are tiny. The smallest of these is Rhode Island which, at 1,545 square miles, is comparable in size to Essex. It took us less than two days to ride through Rhode Island and on our third day in New England we arrived in our third state, Massachusetts. I couldn't resist visiting the 'American Plymouth' and see the Plymouth Rock: incorrectly hailed as the landing place of the Pilgrims in 1620.

The rock itself is decidedly underwhelming, an unattractive lump of granite that's been split, cemented back together and moved back and forth numerous times over the years. It's even more underwhelming when you delve into the controversy that surrounds the rock and its provenance. Aside from the inconvenient yet established fact that the Pilgrims first made landfall on Cape Cod, the first mention of the rock as the landing place in Plymouth is in 1741, when a ninety-four-year-old church elder started a rumour that has somehow persisted for over 250 years. The official website for the town of Plymouth, Massachusetts, is suitably coy about the lineage of the rock, which it peculiarly describes as a 'simple glacial erratic boulder... viewed by more than one million visitors each year'. It's worth seeing to see what all the fuss is about, but don't expect to make much of a fuss yourself.

Our next stop was Boston, home to both Harvard and MIT

Universities, the iconic Fenway Park, the pub from Cheers and an omnipresent Irish charm. It certainly charmed us. We spent two nights in the Jamacia Plain neighbourhood with hosts Tyler and Mary Alice. On our rest day we made the usual city mistake of walking miles and miles without realising it, exploring the university grounds and the city's scenic 'Harborwalk', a linear park along Boston's shoreline. By late afternoon we were shattered and rewarded ourselves with a few illegal brown paper bag beers in one of Boston's leafy parks.

Unlike in booze-loving Britain, in America you generally aren't allowed to drink in public. Some states have arcane laws and you can't even carry your booze from the shop to your car without bagging it. There does, however, seem to be a tacit understanding between the police and the public that placing your booze in a brown paper bag means it is no longer alcoholic and therefore fit for public consumption. Thus, should you buy a few cans from a liquor store (an American 'off-licence') you will generally be offered a few brown paper 'koozies' to accompany your beverage. It's a bizarre system, but it seems to work. We had successfully tested it several times on our trip by this point, and it worked for us again that day in Boston.

Slightly tipsy, we found a local Cuban restaurant and gorged on the tasty combination of rice, beans and plantain. Without intending to, we rounded off the day with a blowout in the most quintessential Irish pub we could find. Our single day in Boston seemed like a holiday and we felt like we'd done the city justice.

***

Almost 11,000 miles into our trip we hadn't met a single framebuilder. On the day we left Boston we met two framebuilders in thirty minutes. We were brewing a coffee at the start of the Bike to the Sea bike path when the fabulously moustachioed Bryan Hollingsworth rolled by. Bryan was riding a stunning steel bike that was obviously custom built. His handmade British racing green Royal H bike was adorned with opulent lugs and a neatly integrated seat clamp. I noticed Bryan was rocking some old school Eggbeater pedals and we lost ourselves down a wormhole of nerdy cycling chat.

We parted ways with Bryan and a few miles later, we were joined on the Bike to the Sea bike path by Jay Borden, crafter of Roulez Cycles. When Jay told us he was a framebuilder we couldn't believe it. We told him about our encounter with Bryan, and it turned out they were friends. Like Bryan, Jay favours a classic look and his own 'Black Betty' bicycle was a beautiful no-nonsense machine, uncompromisingly finished off in black powder paint. Jay noticed a horrible noise emanating from Ties' brakes and said we should follow him to his workshop, which was also fortuitously on our route. He fixed the problem easily, realigning the pads, and we plugged on.

Twenty miles after Jay's workshop we were in Salem, a town infamous for the witch trials that took place in the late seventeenth century. The trials are well commemorated throughout the town in thought-provoking sculptures and exhibits. We discovered that when two young girls displayed convulsions deemed 'beyond the power of epileptic fits or natural disease', the deeply religious community were fearful and looked for an Earthly scapegoat. A literal witch-hunt ensued and over the course of a year 200 people were eventually accused. Nineteen people were executed by hanging and one was pressed to death for refusing to plead.

After our sobering visit to Salem we carried on to Salisbury Beach where we'd arranged to stay with our host Matt, who lived with his dog in a small apartment overlooking the beach. Matt explained to us that as a busy nurse and photographer he didn't have many opportunities to socialise and therefore when he wasn't working, he liked to host cyclists.

He took us to a restaurant in nearby Newburyport and treated us to a slap-up meal. Matt was incredibly open and said he saw Warmshowers as a reciprocal arrangement where everyone benefited. As Matt explained, it gave him the opportunity to go out for dinner and hear interesting stories from a huge variety of people. And his guests, like us, received free accommodation and a scrumptious meal in return. We always suspected this was some people's motivation for hosting, but it was refreshing to have an open conversation about it with someone so comfortable that they brought it up.

Within three miles of leaving Matt's apartment we were in New Hampshire. And twenty miles later we were in Maine. It's not like New Hampshire's tiny (its bigger than both Connecticut and New Jersey) but the narrow Seacoast Region, where the state meets the Atlantic Ocean, certainly is.

The other intriguing thing about New Hampshire is the state motto. As we rode across the Massachusetts-New Hampshire border, we were surprised to be greeted by a sign that said 'Welcome To New Hampshire, Live Free or Die'. We were fond of these 'Welcome To...' state border signs. Here are some examples of signs that raised a smile:

- Alabama – *Sweet Home Alabama*
- Arizona – *The Grand Canyon State*
- Georgia – *We're glad Georgia's on your mind*
- Delaware – *Endless Discoveries*
- Mississippi – *Birthplace of America's Music*

The motto on New Hampshire's sign is decidedly less rosy than those above. It originates from a toast made in 1809 by the state's most famous soldier of the American Revolutionary War, John Stark. While the motto seems a bit morose it would have been worse if they hadn't shortened it; the original was 'Live free or die: Death is not the worst of evils'.

In 1971 the state caused some consternation by mandating that all number plates replace the arguably more appropriate motto 'Scenic' with the melancholic 'Live Free or Die'. This led one resident, George Maynard, to take the state to the US Supreme Court. Maynard was a Jehovah Witness and argued:

"By religious training and belief, I believe my 'government' –

Jehovah's Kingdom – offers everlasting life. It would be contrary to that belief to give up my life for the state, even if it meant living in bondage."

The US Supreme Court agreed with Maynard, noting precedent had been set in a 1940s case where Jehovah Witness children had refused to salute the American flag in public school. I can empathise with the schoolchildren. I wouldn't want to salute a flag that originates from Sunderland either.

CHAPTER SIXTEEN

LOONEY TUNES

12th April – 3rd May 2017
Distance ridden this chapter: 934 miles / 1,503 kilometres
Total distance ridden: 11,919 miles / 19,182 kilometres

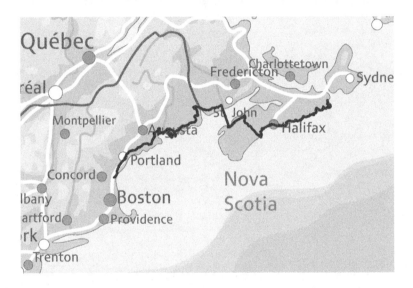

'Welcome To Maine, the way life should be' was the greeting we received from the twenty-sixth and final US state we rode through. There was something poetic in that welcome, as if there was a reason Maine was the last state we'd cycle through. Whilst I can't claim to have found the meaning of life in Maine, or to understand why life should be as it is in Maine, it was a beautiful state that delivered on scenic rides and wonderful encounters with friends old and new. A fitting end to the US leg of our adventure.

Within a few miles of crossing the state border with New Hampshire we were reunited with Ted, who had hosted us over

3,000 miles away in Port St Joe, Florida. We'd stayed in touch with Ted and were fortunate to catch him in his hometown of Kittery. Ted's Maine house was just as impressive as his shorefront pad in Florida, and we caught up on our East Coast adventure and his Iron Man training, over a hearty steak dinner.

Having enjoyed our mini holiday so much in Boston, we decided to stop off in Portland for a few nights, our next stop after Ted's. I was disappointed to miss out on Portland, Oregon, and a few people had recommended its less famous namesake in Maine. Uncle Mo had also given us an extraordinarily generous contribution towards our trip, so we were feeling flush. Nonetheless, we wanted to stretch Uncle Mo's gift and booked a worryingly cheap Airbnb, which turned out to be down a dark alley in an ominously gritty part of the city. We needn't have worried though as the whole place turned out to be quiet, super friendly, and lots of fun.

We also discovered that Portland has a pleasingly high brewery density. After exploring the city's Victorian architecture and modern harbourside, we set about indulging in some flights of beer from the local taprooms. Despite how it may seem from this book, we didn't drink regularly on our trip, and when we did we usually forgot this, inevitably resulting in a cheap night out and an early finish. This was one of those occasions, as we succumbed to the tempting beers – names invariably suffixed with words like 'triple' and 'imperial' – that were far stronger than two cycle tourers could handle.

About ten miles north of Portland we passed an enormous Garmin sign, the company that made the GPS device that navigated us on our trip. I don't know what I expected to say to them, but I uncharacteristically took us off route and down the access road to their office. The security guard didn't seem that confused to see us but explained that the office was closed because it was Good Friday. As we were talking to the security guard we spotted an enormous globe in the background. The security guard explained that it held the Guinness World Record for 'biggest rotating globe' and that it was a bit of a tourist attraction. At the time of writing, 151 people have ranked the globe number one on Tripadvisor's list of nineteen 'things to do' in Yarmouth, Maine. At the time of visiting, we had no idea about the globe and we explained to the security guard that we'd taken a short detour on our long trip because we spotted the Garmin logo.

The security guard enjoyed our story and said although the office was closed for Good Friday we were welcome to come in, with our bikes, to have a closer look and take a photo. We took him up on the offer and the closer we got to the globe the more obvious it was why it was such an attraction. I'd never wondered how big the world's biggest rotating globe would be, but this thing had a whopping 129 feet circumference and weighed almost three tons.

***

The last time we'd seen Brent he was scavenging for smouldering embers at San Elijo State Park, California, in November 2016. Brent had then enthusiastically told us that he was doing everything he could to land a job working as a deckhand on a schooner in Maine for the 2017 season. We weren't surprised to hear Brent secured the gig and we were delighted that he offered us a night on the schooner, which was being prepped for the season at its dock in Camden, Maine.

Launched in 1871, the Schooner Lewis R. French is the oldest commercial sailing vessel in the US. Well-heeled yachties pay serious money to spend a few nights aboard the handsome vessel. When we arrived the entire hull of the schooner was wrapped in white vinyl. Brent helped us pull our bikes aboard, being careful not to disturb the vinyl cloak too much, and then cooked a whole lobster in the galley kitchen. The lobster was delicious and as we complemented the chef Brent confessed he'd never cooked lobster before, but needed to start learning as he'd soon be churning out daily servings of the crustacean for paying guests with high expectations.

Catching up with Brent, the cycling MacGyver

We had a fun night trading stories about our respective adventures over the past five months and then retired to our cabin below deck. The following morning Brent practiced his breakfast skills on us and then rode with us as far as Belfast. Once self-proclaimed the 'Broiler Capital of the World', Belfast's many 'chicken factories' processed an average of 250,000 chickens a day. Modern-day Belfast gives few clues of its once-proud industrial past, and the poultry processing plants have given way to a genteel waterfront town, which provided a picturesque setting for a spot of lunch.

After a quick pitstop at Belfast, Brent turned back on himself and headed back to Camden to continue prepping the fancy schooner. We were taking a detour to Castine – a recommendation from Ted in Kittery – where we had arranged to stay with hosts Barbara and Jarlath. The ride there was spectacular. Ted had waxed lyrical about Penobscot Bay, the body of water that separates Belfast from Castine. He had assured us that the detour to the cute town of Castine would be worthwhile, and he was right. As soon as we left the main road at the small town of Orland there descended a quiet serenity which we hadn't experienced anywhere else on the highly populated East Coast. Every shade of green was represented in the trees which lined the road and occasionally parted to allow a view of the calm Penobscot River. It wasn't dramatic scenery but we had it all to ourselves and that was soothing.

When we arrived in Castine, Barbara explained that her family had plans that evening but they'd prepared our room and a DIY dinner. She showed us to the kitchen and to our delight she'd laid out all the ingredients and accoutrements for us to make dozens of Vietnamese summer rolls. We were salivating and waited for the family to leave before frenetically assembling and devouring scores of the rolls, a dish that we'd go onto create budget versions of throughout the remainder of the trip.

***

Acadia National Park was our next stop after Castine. Bar Harbor is an old colonial fishing village and now the gateway to the Acadia National Park. It is also famously the end (or start) to both the ACA's Atlantic Coast and Northern Tier cycling routes. It's one of those 'must see' places that everyone assumed we'd go to.

We arrived on 17th April 2017. This was well out of season and we reaped the rewards with a stay at the KOA campground with our host, Barbie. Acadia National Park itself is located on Mount Desert Island, with small sections also located on the seldom visited Schoodic Peninsula and Isle au Haut. The park attracts three million visitors per year during peak season between May and October, who pay to drive their vehicles around well signposted loop roads. The highlight is the road up Cadillac Mountain which is so busy at peak times that you need to reserve a time slot.

In April, on a bicycle, park rules are decidedly more *laissez-*

faire. You can go pretty much anywhere and it doesn't cost you a penny. As snow fluttered around us, we rode the twenty-seven-mile Park Loop Road and took the detour up to the top of Cadillac Mountain. It's easy to see why it's so popular. The island is pimpled with small mountains that are covered in intensely green trees. Everywhere you look you can see crystal clear water, whether it's the shoreline, a harbour, a pond or a lake. Majestic birds fly overhead – we saw peregrine falcons, eagles and osprey – and there are hidden beaches everywhere.

Barbie had given us loads of recommendations for our route, which made sure we didn't miss anything and made the whole experience even more rewarding. She had also highly recommended we ride the Schoodic Peninsula loop road, saying that it's always quieter than Mount Desert Island and at this time of the year we'd have the place to ourselves.

I found a 3-star guest house online called the Bluff House Inn at the start of the Schoodic Peninsula. We decided to ride there, ditch our bags and then ride the scenic loop road unburdened. When we arrived at the Bluff House Inn I thought we were at the wrong place. It was stunning, resembling a grand teak cabin, wrapped in a veranda strategically positioned to give a perfect view over Frenchman's Bay and its collection of tiny islands. We hadn't stayed in a motel for a while and I was feeling more sheepish than usual about clip-clopping into reception on my cycling cleats, clad in my merino tights and dank cycling jacket. The receptionist was typically unfazed and showed us to our room.

I was excited about riding the loop but having seen the veranda Ties wasn't going anywhere. Whilst Ties made the most of our luxurious '3-star' guest house, I trundled down to the tiny town of Winter Harbour and rode the deathly quiet and incredibly scenic Schoodic Loop Road. When I returned to the guest house I found Ties in the lounge, chatting to a Polish family who were on holiday exploring New England. The family had prepared a huge spread of cold meats and cheese, which they'd sourced from specialist delis in Boston, and invited Ties to join them. I was emphatically greeted by the gregarious matriarch of the family and promptly presented with a delicious plate of unfamiliar treats.

***

We waved goodbye to the US at Calais, which is named after the ferry port in France, honouring the support that France gave to the US during the American Revolutionary War. Just like our experience at the Minnesota-Ontario border, we breezed through Passport Control with best wishes and smiles from the bubbly Canadian Border Services Officers. Our time in the US was hugely enjoyable. We were bowled over by the phenomenal variety of the country's landscape, surprised at how safe we felt cycling on most of the roads and humbled by the kindness we experienced from the vast majority of strangers. However, the bureaucracy in the US was mindboggling and our visa debacle had caused so much unnecessary stress that I was pleased to be back in a country that didn't seem to place any restrictions or administrative requirements for our stay. It was also relaxing not to have the spectre of politics hanging in the air or to worry about inadvertently offending any covert Trump supporters.

We had ridden through the US at an extraordinary time in its political history. We were there for the 2016 presidential primaries and caucuses (essentially, meetings where local supporters register their preference for presidential nominees), the election campaigns, Trump's inauguration and the backlash that soon followed. We left the US on 20th April 2017, just days before Trump completed his first 100 days in power. It's a ten-month period that will probably go down as one of the strangest in US history, led by undoubtedly the most bizarre US President of all time.

***

Our third stint in Canada started in New Brunswick, on our

old friend the Trans-Canada Highway. This was an especially quiet stretch that was flanked by imposing rock walls decorated with pine trees and the odd goat here and there.

Up to this point of our trip we hadn't been sure when we'd finish. The northeast of Canada had plenty of interesting areas to explore and we were only restricted by our budget. However, now we had a hard deadline; our friends Matt and Dipti had scheduled their wedding for 22nd July 2017 back home and we weren't prepared to miss that. This still gave us plenty of time to explore Nova Scotia and Prince Edward Island (PEI), before returning to Toronto via Quebec. I mapped out a promising route, starting with a ferry across the Bay of Fundy to Digby, Nova Scotia.

The closer we travelled to Nova Scotia, the more people eulogised about it. There was a genuine excitement in people's eyes when they talked about the 'Scottish' landscape and the 'world's friendliest people'. This was always caveated with a word of warning about the weather and a scepticism that we weren't going at the right time of the year. The popular joke was that Nova Scotia has two seasons; winter and summer, the latter of which lasts six weeks from early July to mid-August. Ties was nervous. She doesn't enjoy the cold and didn't like the sound of riding into winter. I shrugged off the warnings. I shouldn't have done, because her apprehensions proved well-founded and I still bear the consequences today.

On our second night in Nova Scotia we couldn't find anywhere to stay. This was something else we'd been warned about – and something else that I'd neglected to take seriously – as the province's short summer also meant the tourist season started late and most campgrounds and motels were still closed. After sixty miles we decided to call it a day. We'd ridden eleven days without a break and were fifty-odd miles from our next Warmshowers hosts. We didn't have the energy to keep going any further and we found a layby just off Highway 10, near the community of Albany. Within minutes of setting up camp an enormous pick-up pulled up and a man shouted "You can't camp *there*..."

Spirits were low and energy levels non-existent. Our hearts sank. The man stepped out of the pick-up, revealing his tree-like stature, and finished his sentence by adding "...because I know an *amazing* camping spot that I'll take you to."

Those few words had made us incredibly happy and Cal, our saviour colossus, was fast becoming our favourite person. He stood patiently while we took our half-pitched tent down and then he lifted our kit into his flatbed. As we drove down the sandy road Cal explained that he was taking us to his favourite fishing spot and he'd be camping there himself if he wasn't doing a night shift. When we arrived at the secret hideaway we could see why Cal had been so enthused. There was a flat, sandy patch right on the edge of a pond-still lake completely surrounded by aromatic pine trees whose reflection was mirrored in the water, with beavers splashing about near the water's edge. It was the quintessential view of Canada and the dream wild camping spot that we'd never have found had it not been for Cal.

For the first time in a while, I slung our rope over a tree and made a bear locker. On this occasion less for the threat of bears and more for the nuisance of racoons and coyotes, which are common in Nova Scotia. We turned in that night noticing two things: firstly, the call of the common loon, which is more wolflike than avian. It is a haunting but beautiful wail, that is very hard to describe: I recommend Googling it if you've never heard a common loon. The second thing we noticed was it was cold. Bitterly cold. Ties had chosen to listen to her body and had gone to sleep wearing every piece of clothing she owned. I chose to ignore the cold and grin and bear it in my UK-rated three-season sleeping bag. After finally shivering myself to sleep I woke up in the early hours with a strange hot itch in three of my toes. Days later I would discover that this unusual feeling was in fact chilblains, and four years later I now experience the same strange hot itch, in those same three toes, whenever the mercury goes below ten degrees Celsius.

Cal's wild camping spot with our bear locker system

That morning we emerged from our tent to find it covered in frost, the first time that had happened. There was also a shopping bag outside our tent. At first we assumed it was rubbish but when we opened it we found an old milk bottle filled with water, four apples and a handwritten note from Cal that simply said 'Hope it wasn't too cold last night. Some water and apples for you'. Cal: a true gentle giant with a heart of gold.

We followed the LaHave River until it flowed into the Atlantic Ocean at Riverport, home of our hosts Mary and Bernie. Mary had said she'd be delighted to host us for two nights and we were looking forward to a rest day. Our hosts had a beautiful home with a farmhouse-like kitchen dominated by a hefty block table. They told us all about their homegrown vegetables and prepared some exceptional vegetarian meals with their organic produce. Mary and Bernie could sense we wanted to chill out and they left us to loaf around the next morning before taking us for a scenic drive down to Fort Point Lighthouse and Rissers

Beach Provincial Park. We had a wonderfully relaxing two nights with our hosts and sensed that this was their general retirement vibe. Seemed idyllic.

Yet again, the night after we stayed with Mary and Bernie we struggled to find any accommodation. Eventually we spotted a campground at the small community of Hubbards. We found the owner and optimistically asked for a pitch, only to be told, again, that we were too early and the campground wasn't open. We'd clocked a picnic area just before the campground and we asked the owner if he thought anyone would mind if we spent the night there. He ruminated for a bit and then said that while his campground wasn't open, we could sleep in one of his rental cabins for free to stay out of the cold. We thanked him profusely and he called for someone to escort us. The cabin looked like a proper lumberjacks' home. With its exposed wooden walls and log-burning stove it was incredibly authentic and the lady who showed us round even turned on all the utilities so we could shower, cook and stay warm.

We tried finding the owner the next morning to thank him again but there was no-one to be found. Staggered once again by the level of trust placed in complete strangers, we pedalled on, happy in the knowledge that we had only a short ride to Halifax, where yet another Warmshowers host had agreed to put us up. On the way, my bike developed a frustrating mechanical issue; the belt kept coming off when changing gear. We were almost 12,000 miles into our trip and until now my belt hadn't come off once. Knowing I was unlikely to find a replacement belt easily, I'd taken a spare with me from London. After the belt had come off a few times I put the new one on, but this didn't solve the issue.

When we reached Halifax I found a decent bike shop, and to my utter joy and amazement the mechanics seemed unfazed by my unusual setup, but declared my chainring and cog (sprocket) screwed. They advised that belted drivetrains wear the same way as their chained counterparts, with the teeth becoming rounded and the belt therefore slipping off. The good news was that shop was a dealer of the company who made my belt. The even better news came when they offered to order the parts for me and then post them to wherever I thought I'd be a few days after they arrived at the shop. In the meantime, they pulled my rear wheel back in the dropouts as far as possible to keep the belt on with tension.

As it turned out, this fix had to last the remainder of the trip. Despite me successfully picking up the replacement chainring and cog two weeks later, none of the umpteen bike shops I visited could work out how to wrestle the cog off of my bike. The 'belt-tensioning' workaround kept the chain on but also somehow stopped me accessing gears twelve to fourteen. This gave me a maximum flat surface speed of fifteen miles per hour and meant I couldn't keep up with Ties when there was no gradient.

On reflection, I'm glad I chose to run a belt on my bike and I'll (probably) do it again on our next tour. Although it's nigh on impossible to find and fit replacement parts on the road, 11,700 miles is a good innings. By comparison, in that same distance we had to replace Ties' chain six times due to premature elongation, or 'chain stretch' as its commonly called. Since our trip, the company who made my belt has also changed the fitting of the cogs so you no longer need such a rare tool to replace your knackered cog.

After our exceptional service at the bike shop we had a short ride to stay with our host Pascal and his family. Pascal worked as a guide for a cycle touring company, which had allowed him to undertake two-wheeled travel across Canada, several countries in Europe and his favourite place, Cuba. He yearned for the road and insisted on escorting us out of Halifax and onwards. Pascal was eager to take us on a stretch of the Trans Canada Trail that intersected Cole Harbour. We were reluctant, still traumatised by battling boulders and ATVs back in BC. He assured us it wasn't like that and he was right: the bike path was perched atop a causeway that stretched out into a beautiful harbour surrounded by pine trees. The surface was hard packed and the views were spectacular, as was the surreal experience of riding through the middle of a capacious body of water.

Pascal turned around at Seaforth and we carried on riding over the constant lakes and harbours that interrupt Nova Scotia's coastline. We'd been warned that this stretch of the coast would have no accommodation at all and after my chilblains incident at 'Cal's lake' I decided to take action. Before we left Halifax we popped into MEC (an upmarket Go Outdoors) and bought some 'Canadian weather' sleeping bags. There was a post office next door and we mailed our 'UK weather' sleeping bags, along with some other non-essential belongings, to our friends Greg and Michelle in Toronto. This proved to be a shrewd move as we spent the next three nights roughing it in various roadside locations.

The weather was growing increasingly inclement and the precipitation combined with lack of facilities was making it difficult to enjoy what was otherwise a picturesque stretch. After our third night of wild camping, we were excited to be staying with Warmshowers host Grayson. The weather had other ideas though, inducing Ties' third and final meltdown of the trip.

We'd spent the previous night sheltered from the worst of the weather under the overhang of the Goldboro Interpretive Centre. We packed up and left in the freezing rain, pleased to know we only had forty miles to Grayson's. After two hours we'd managed a pathetic twelve miles. The rain was horizontal and the wind was testing the limits of what's possible on a bike. Ties was hurling inflammatory insults into the Baltic headwind, doing her best to turn the pedals and managing six miles per hour on the flat. Through the dense rain she spotted a sign that read 'Lonely Rock – Seaside Bungalows – 150m' and without saying anything turned off the road and followed the sign until she found the lonely bungalows. Despite it being 9:30am the proprietor took pity on us and let us check in. The owner pointed out that we'd timed our arrival well as they'd only opened for the season the previous day. I was not overly enthusiastic about our stay – it was the most expensive place we'd stay in Canada (up to that point) and Grayson's was only twenty-seven miles away – but I wasn't going to win this one and took solace in the indulgence of showering for the first time in four days.

Grayson was also philosophical about the delay. I messaged him to apologise for our late arrival and his laid-back response was "Not a problem Chris. It's your trip not mine. You are more than welcome tomorrow as well" adding "If you get to the house before I get home let yourselves in."

The weather was decidedly less emotional the next day and we did indeed arrive at Grayson's before he did. When Grayson turned up we were excited to discover that not only was he as friendly as his communication suggested, but he was also a bona fide Royal Canadian Mounted Police (RCMP) Officer, or a Mountie, to you and I.

Hitherto my only exposure to Mounties was through the cheesy 1990's TV series *Due South* and the Mountie-themed tat that graces every Canadian themed gift store. It was a revelation to hear that the RCMP are the federal police service of Canada and are responsible for policing misdemeanours at the spikier end of the law, like organised crime and drug trafficking. Furthermore, and contrary to all the gift store paraphernalia, they no longer ride horses on a routine basis. This last bit was mildly disappointing.

We had a great evening with Grayson hearing about his RCMP work in the remote Yukon territories and his motorcycle tours of South America and Africa. Grayson was excited for us that we were heading to Cape Breton Island to ride the Cabot Trail, but warned us to be careful as there was still 'some snow'. As we left the next morning Grayson insisted on giving us his mobile number and said not to hesitate to contact him if we ran into any trouble. We thanked him for his kind offer and pedalled off into a light but icy wind, eagerly anticipating the wild, remote landscapes promised on Cape Breton Island.

CHAPTER SEVENTEEN

BIG POTATOES

4th – 31st May 2017
Distance ridden this chapter: 1,147 miles / 1,846 kilometres
Total distance ridden: 13,066 miles / 21,028 kilometres

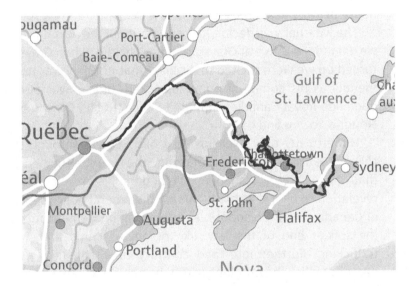

Our route around Nova Scotia was predetermined by the ride to Cape Breton Island, home of the Cabot Trail. It's impossible to overstate how excited I was about riding the Cabot Trail, routinely voted one of the world's best cycling routes. Cape Breton Island itself is oft ranked amongst the best islands in the world and we were intrigued to experience the mix of Gaelic and French culture.

Ties was nervous about the Cabot Trail. I didn't sugar-coat the elevation profile, which includes two monster climbs and a total of 14,000 feet of ascending in 147 miles. This is what I would call 'Dartmoor hilly', and far hillier than anything we'd ridden to date in North America. We planned on short days and had no idea where we'd spend the nights, assuming we'd wild camp in rest areas.

The night before we started the Cabot Trail we ended a hilly day's ride at Margaree Forks, where we pitched our tent round the back of the visitor centre, conveniently located on the Cabot Trail Road. We always made a swift exit when we weren't exactly sure if we should be camping somewhere and the next morning we were on the road by 7am.

Ten miles into the Cabot Trail we reached the coast and were greeted with a sign welcoming us to 'Region Acadienne'. We'd entered a part of Cape Breton Island that speaks Acadian French, a variety of the language that retains characteristics from sixteenth century French. This was our first encounter with the many varieties of French spoken in Canada's Atlantic provinces, which proved more challenging than originally anticipated.

As we rode though the tiny Acadian community of Terre Noire, the language was the least of our problems. The wind had gone from breezy to bonkers and we were struggling to stay upright. I was adamant we should push on. The wind may have ratcheted up, but so had the scenery: the view in front of us was what all the Cabot Trail promotional material had promised and the road itself was perfectly tarmacked and devoid of any traffic at all. However, as we approached Saint-Joseph-du-Moine, three miles later, I had to concede that the wind was verging on hazardous. We peeled off the road and took shelter at the side of a garage. I was longing for the wind to stop so we could drink in the spectacular coastal vistas that we'd seen from our sneak preview of the Cabot Trail. Ties on the other hand had sensibly already ended the day in her head and was rightly pointing out the absurdity of cycling in wind that we couldn't even stand up in.

After clinging onto the side of the garage for thirty minutes an elderly gentleman approached us and started speaking to us in what we assumed was Acadian French. We deduced that he was called Hector and it was evident from his frequent pointing at the sky, wagging of his finger and tutting that he didn't think we should be cycling. Hector trundled off and returned in a small pick-up, followed by his wife Marie in her car. Through the language of charades Hector made it clear that he and Marie were going to transport us to the nearest motel where we should then remain put for the foreseeable. We thanked the lovely couple *en français* and bundled ourselves, our steeds and our kit into their vehicles.

The Cabot Trail through Terre Noire

Within ten minutes we arrived at the fishing village Chéticamp. We tried two motels but they hadn't opened for the season yet. At the third motel the owner, who was Acadian but spoke excellent English, explained that he was a lobster buyer and that the wind had delayed the start of the lobster season. With no lobster to purchase, he was considering opening the motel that day and our request for a room had made his mind up. As we installed in the room, it was clear that the animated conversation between the owner and Hector was one that would be retold to a few more people. We thanked Hector and Marie profusely and we both understood their response as they continually muttered *de rien*.

For the second time in five days our ride had ended at 9:30am. I was frustrated but inclined to concede that weather capable of preventing lucrative lobster boats from setting sail is not conducive to a pleasurable ride along the coast. I was even more phlegmatic about it when I discovered the wind had a name – *les suêtes* – whose etymology originates from Chéticamp and is Acadian for strong south-easterly winds.

Les suêtes are particular to the west coast of Cape Breton Island and have been recorded in excess of 120mph, equivalent to a category three hurricane. One week after we experienced *les suêtes,* they returned, overturning an articulated lorry in Saint-Joseph-du-Moine. The motel owner also told us that a few months before we arrived, *les suêtes* had snapped a wind turbine in half in the nearby village of Grand Étang. He also added that the lobster boats weren't planning on venturing out the next day, which was a sure sign that *les suêtes* were hanging around. We didn't need to hear anything further and extended our stay at the motel.

After two nights in the motel there was still no sign of *les suêtes* moving on. With a heavy heart I parked my pride and my aspirations for riding around the Cabot Trail and reached out to Grayson. Grayson wasn't surprised to hear from us and told us to hold tight while he picked up his mate's trailer. When Grayson arrived he told us that he had no plans for the day and offered to drive us around the Cabot Trail Road that we'd planned on spending three to four days riding. We were over the moon and quicky accepted his offer.

The views were everything I'd hoped for. Endless miles of

rugged yet verdant coastline, only interrupted by the snaking Cabot Trail Road that looked like a glossy ribbon wrapped around the shore. As Grayson drove up the two steep categorised climbs – French Mountain and North Mountain – I could feel Ties' stare piercing the back of my head and was momentarily relieved that we hadn't ridden them. At the top of North Mountain we saw the snow that Grayson had warned us about. He wasn't kidding, there was piles of the stuff and I contemplated how cold it would have been up there in the tent.

Me climbing on the snow at the top of North Mountain

The whole loop was spectacular and Grayson had some more surprises for us on the way back to his house. He insisted we pick up some beers from the organic Big Spruce Brewing – just off the Cabot Trail Road – and then took us to Charlene's, an unassuming café in the rural community of Whycocomagh, highly regarded for its seafood chowder. We all had a good fill at Charlene's and then polished off our delicious organic beers at Grayson's.

***

Nova Scotia was extremely challenging. The weather had been particularly nasty, but the lack of campgrounds and bike issues had also contributed to a tough fortnight. As we left Nova Scotia on the ferry from Caribou, we had everything crossed that our time on PEI would be less problematic.

PEI is the smallest of the ten Canadian provinces and at 2,185 square miles is ten times smaller than the next smallest province (Nova Scotia). It also has the highest population density of all ten Canadian provinces, but don't think its crowded: at twenty-five people per square kilometre, it has the same population density as Powys, the most sparsely populated area in Wales.

PEI is famous for one thing; potatoes. It grows twenty-five percent of Canada's potatoes, which is impressive considering it constitutes less than 0.1% of its land area. Everywhere you turn in PEI there are reminders of the importance of the starchy root vegetable. We couldn't resist taking a photo of the sign directing visitors to the Canadian Potato Museum and we quickly grew bored of playing the 'first-one-to-see-the-next-potato-farm-wins' game. With all this potato growing you'd think the scenery is pedestrian, but it isn't. The potatoes are harvested in red soil which looks fantastic with the patchwork farms punctuating the otherwise lush green island. The sandstone cliffs are also red and there's an immaculate bike path – the Gulf Shore Parkway – that follows the unusual crimson shoreline.

As with Nova Scotia we did have problems findings places to stay. On our first night we sheepishly camped in the closed PEI National Park Stanhope Campground. Some locals told us it was fine for one night and said that the facilitates, including the showers, would all be available. They were right and we felt a bit cheeky but it was that or a potato farm.

The Gulf Shore Parkway

On the second night we were riding down what turned out to be a no through road when we saw an elderly couple doing a roadside litter pick. We said hello and the unexpected response was "What in the hell are you doing?"

We briefly explained our trip and within minutes Winston and Edith had invited us to stay for the night. We had a wonderful evening, plied with hearty home cooking and stories from the couple who between them had lived on PEI for more years than Canada is old.

After devouring a mountain of Edith's glorious pancakes, the next day we hopped onto the North Cape Coastal Drive, determined to see the landmark North Cape Lighthouse, but wary of the strong winds after our Cape Breton Island experience. Sure enough, we had to battle strong crosswinds and freezing rain. Severe winds were forecast for the next day and we had a foreboding sense of déjà vu. After seventy-four blustery miles we found a motel near the settlement of Christopher Cross. It seemed ironic to us that we ended our day at a place who's namesake penned the hit song 'Ride Like the Wind', but I'd finally learned my lesson about the perils of North American weather. We hunkered down in said motel and hibernated from the elements for two nights.

We left the motel under a blanket of mist and freezing temperatures, wondering where we'd sleep that night but looking forward to seeing the West Cape. Five miles in we stopped at a grocery store to refuel and within minutes a local man called Glen had invited us to stay. We gratefully accepted and plugged Glen's address into the Garmin, delighted to see it was en route.

Frustratingly, our forty-mile ride turned into sixty miles due to an unexpected detour. Flooding had caused a cross section of the coastal road to sink. The local authorities had decided to leave the road open and draw attention to the enormous dip with red X's spray-painted on the road. An unwary driver hadn't noticed the warning signals and had driven straight into the dip, taking a nasty hit to the chest. This unfortunate accident had convinced the authorities that maybe they ought to close the road and fix it after all, although not, alas, quickly enough to prevent our twenty-mile recalibration.

Eventually we found Glen's house and were warmly greeted by his family and impossibly cute kitten. Glen was also hosting a Welsh couple, who were travelling around PEI delivering sermons at Gospel Halls. We all feasted on an amazing home cooked meal and then joined the family for the sombre service.

After falling prey to so many gruelling winds, the next day we felt the full benefit of the prevailing north westerlies and were blown down Highway 2 towards Summerside. By noon we'd cycled almost forty miles and took an early lunch in the oceanfront park. Within no time we bumped into Susan, a local Warmshowers host. Susan was walking her dog, killing time while waiting for her friends Jane and Dennis to return from a training ride for their upcoming Trans-Canada cycle tour. Susan joked how they'd love to have met us and moments later we ran into Jane and Dennis as they were returning from their training ride. Within no time they offered us a place to stay in return for kit and route tips. We gladly accepted the invitation and spent a wonderful afternoon and evening in the couple's unique house, which they'd converted from a boatshed.

From Summerside we had a short but scenic ride down to the Confederation Bridge, which joins PEI with New Brunswick. Sadly you can't cycle over the epic eight-mile-long construction, so you have to pay to be shuttled over in a pick-up. The truck dropped us off at Cape Jourimain Nature Centre and moments later we were swerving debris from a car that had hit a moose, which was now graphically displayed in the shoulder of the highway.

We started our second stint in New Brunswick following the Acadian Coastal Drive. Our day ended in Cap-Pelé where we stayed with a Dutch couple Philippa and Robert. Ties met Philippa on the Facebook group 'Dutch Women abroad' and we enjoyed a days' rest, eating delicious food, hanging out with new friends and haring around on our hosts' John Deere ride-on lawnmower.

***

The language in Eastern Canada kept us on our toes. Nova Scotia and PEI are officially English-speaking, with dispersed pockets of Acadian French, Canadian Gaelic (derived from Scottish Gaelic) and Mi'kmaq (a First Nation language). Quebec is the only officially French speaking province, with seventy-eighty percent of Québécois speaking Quebec French. However, we found New Brunswick the most challenging province in Canada, as it's the only one that is officially bilingual and we never knew how to start a conversation.

Our route north through New Brunswick was made tough by the unwanted presence of a stiff headwind. We had arranged to stay with Warmshowers host James in Miramichi, but after six hours that day we'd managed just over forty miles and we knew we wouldn't make it. We persevered through the wind for another ten miles to the campground at Kouchibouguac National Park. One of the Park Rangers persuaded us to take the bike trail north the next day, assuring us it was in great condition. It was an incredibly scenic trail meandering through woods and along the river. However, after a few miles we ran into the aftermath of the previous day's winds, battling constant fallen trees. We abandoned the trail and followed the scenic Highway 117 that hugged the Gulf of Saint Lawrence, wishing we'd followed our instincts from the start.

James had told us it was no problem turning up a day later. As soon as we arrived at his bungalow in Miramichi he explained he had plans, gave us a tour of his place, drawing attention to the swimming pool and beer fridge, and promptly left saying he may see us in the morning if we're late risers. Although several hosts had let us stay in their house unaccompanied it never failed to amaze us how trusting people are and how fortunate we were to experience such generous hospitality.

We followed the main highway out of Miramichi, relieved that we'd ignored everyone's scaremongering to avoid the road, which was practically empty and afforded sweeping views over the dense but low canopies. It also gave us our second, and only other bear sighting of the trip. At first we weren't sure we'd seen a bear, thinking we'd witnessed a rottweiler run onto the road in front of us. However, a pick-up driving south, that narrowly avoided hitting the animal, had stopped to check we'd seen it before we unwittingly got too close to what he confirmed was a small black bear.

***

Campbelltown is a tiny city on the Restigouche River that separates New Brunswick and Quebec. When we arrived in Campbelltown we had an important decision to make about our onward route. I worked out that with the time it would take us to ride back to Toronto, we comfortably had three weeks to spare to catch a flight back for Matt and Dipti's wedding. We narrowed our options down to two choices: take a detour from Campbelltown to ride the Gaspésie, or head straight to Toronto and spend some time in Newfoundland on the way home.

The Gaspésie is a rugged peninsula that stretches from Campbelltown round to Rimouski on the St Lawrence River. It's well known for its stunning coastline, frequent whale sightings (including belugas) and tough gradients. Countless people had recommended we ride the Gaspésie, but had also warned us that we may encounter winds akin to *les suêtes*.

Newfoundland, we were told, offered a truly unique cultural experience, guaranteed whale spotting and icebergs. Although Newfoundland involved the added complication and expense of another flight, we were intrigued by the culture and ultimately sold by the chance to see icebergs. In addition, from Toronto, Newfoundland is on the way home; direct flights from the capital St. John's to London take just five hours.

We booked our flights to St. John's and headed inland through Quebec, following the Matapédia River to the superbly titled Val-Brillant. We continued northwest to Mont-Joli, where we picked up both the St Lawrence River and some gratifying tailwinds. As we were being blown down to Porc-Pic we saw a sign for *la Route verte*, accompanied by *la Littoral basque* motifs. I'm not one for spontaneity but something intrigued me about it and we peeled off the road to follow the route marker. After a mile we came to some strange looking yellow gates that were just wide enough for a bicycle but said *'propriété privée'*. Despite my dodgy French, I could translate that. Confused, and mildly unnerved, we carried on through the gates and onto the *propriété privée*. Our easy, flat ride instantly became a gruelling climb up a twenty percent gravel road. We took a rest and two Québécois ramblers started chatting to us. They spoke great English and explained that four brothers had allowed this section of *la Route verte* to pass through their private land. The brothers had even paid for the infrastructure and had become local heroes as a result. Buoyed on by this heartwarming story we struggled up more hills and eventually came to a lookout built on a suspended platform, entitled *Belvédère Beaulieu*, jutting out above the trees and river below. The majestic view of the St Lawrence River, over the dense forest canopy, was made all the better by the fact hardly anyone else seemed to know about it. The trail meandered its way steeply down from the lookout before pointing skywards up a mean twenty-five percent gradient. It was a short climb but brutal on our substantial steeds. I rode to the top at tortoise pace and, 325 days in, for the first time on the trip, walked back down to ride Ties' bike up. Once the heart rate was back below 200 beats per minute, we re-joined the road and ended the day soon after.

The benevolent tailwind persisted, wafting us all the way to Quebec City, where we took a few days off to recuperate and celebrate our two-year anniversary. It dawned on us as remarkable that we'd not only survived living in each other's pockets for almost one year, but we'd done so after only knowing each other for a similar period of time. We ambled around the European-looking cobbled streets, eating poutine (chips with gravy and cheese curds) – and plotting the final few weeks of our mainland adventure.

CHAPTER EIGHTEEN

COMPLETING THE LOOP

1st – 17th June 2017
Distance ridden this chapter: 734 miles / 1,181 kilometres
Total distance ridden: 13,800 miles / 22,209 kilometres

When I was in London researching bikes and touring equipment, I spent hours flicking through the online Loaded Touring Bike Photo Gallery. For anyone not interested in cycle touring set ups this is probably the most boring website you could possibly find. However, should you be green to the world of cycle touring and eager not to be, it's a treasure trove of information.

I was most interested in panniers, keen to discover an alternative to the ubiquitous bucket system that leaves you digging around for that item that's invariably at the bottom of the bag. There was one brand of bags that intrigued me, Arkel. They have a zip that goes all the way around so you can access everything easily, and they're covered in external pockets and removable pouches. They also have a natty 'tent tube' that attaches to the back of the pannier, meaning your single biggest piece of equipment doesn't take up valuable bag space. I ended up using Arkel panniers on my bike, and the company followed our trip on Instagram. As we approached their head office in Sherbrooke, Arkel contacted us on Instagram and offered us a tour of their workshop and a place to stay for the night.

We spend two days ploughing through a relentless headwind, to make our way from Quebec City to Sherbrooke. En route, we rode through a town bearing the ominous name of Asbestos. We later discovered that the town's name was derived from its proximity to the Jeffrey Mine, which until 2011 was the world's largest asbestos mine. In December 2020, wary of the negative connotations associated with the hazardous mineral, the town administration changed the name to Val-des-Sources. This wasn't without opposition and the name change was nearly kiboshed by a petition to retain the Asbestos moniker. Alas, sense prevailed and the petition was squashed by Quebec's Minister of Municipal Affairs and Housing.

Eventually we beat the headwinds and made it to Arkel. We had a tour of the workshop, amazed by how small it was, but reassured to hear that all of their bags are lovingly handmade in that very building. It was also fascinating to hear about Arkel's outreach work. Since 2003 Arkel has employed local people with learning disabilities and pervasive developmental disorders, to assemble their bags and gain valuable skills in the process.

Yves, who we discovered we'd been messaging via Arkel's Instagram account, invited us to stay with his family for dinner. We were also joined by Arkel's owner Paul and we had an entertaining evening sharing stories and feasting on delicious barbequed food. It was amazing to meet the people behind a company I admire so much and discover that they're so down to earth and genuinely interested in the stories from those who use their products.

Yves and his daughters guided us over the Saint-François River the following morning and we then pointed west for the two-day ride to Montreal. Thankfully, the gruelling headwinds had subsided and we enjoyed an agreeable ride to Canada's second biggest city along well maintained and scenic bike paths. Our final passage onto the island city was via the hulking Jacques Cartier Bridge, undoubtedly the most challenging bridge of our trip. We'd been cautioned about the steel beast but we're partial to a good bridge so shrugged off the warnings. The main issue for us was how narrow the shared cycle and pedestrian path was. Due to our bike's bulbous rear ends, we were constantly forced to stop and move aside to let oncoming riders by. Each time we stopped we then had to get our heavy rigs going again on ascents approaching seven percent. It's also the third busiest bridge in Canada, so not kind to ear drums that aren't safely cocooned within the shell of a motor vehicle.

The man who shares his name with the bridge is a big deal in Montreal. Jacques Cartier is said to have been the first person to travel inland in North America, in 1534. A year later he returned with a fleet and took his smallest vessel down the St Lawrence River to Hochelaga, or Montreal as we know it now. Cartier claimed the land for the Kingdom of France, who kept possession until the British wrestled it away after a lengthy war that ended in 1763.

Our arrival in Montreal coincided with the annual *Tour de l'Île de Montréal*, the Island Tour of Montreal. The event saw tens of thousands of cyclists ride around thirty miles of closed roads on the island city. We arrived too late to benefit from the traffic-free streets but a pervading festival atmosphere remained, and the charming waterfront thronged with hordes of cycling enthusiasts and their bicycles.

We had arranged to meet Eddie, an old work colleague of mine, in Place Jacques-Cartier. Eddie treated us to several varieties of poutine and we devoured them while watching the superyachts manoeuvre around the city's port.

As we polished off our stodgy lunch, Mathieu and Léonie joined us. We had been in touch with the couple on Instagram and they had offered to show us around and let us stay in their apartment. We hit it off immediately, sharing a passion for adventure, cycling and animals. After a fun ride through the city we had a warm greeting from their adorable dog, Zipper. Mathieu was determined to rustle up some salmon tartare and the two of us embarked on a hunter-gatherer mission to the local supermarché.

Mathieu and Léonie showed us around the classic Volkswagen T25 they were readying for a big adventure around the national parks of Canada and the US. We shared our experiences of the few we'd visited and listened with envy at their budding excitement which we recognised from our own pre-travel anticipation.

***

We left Montreal in a torrential downpour, pleased that we only had a twenty mile ride to stay with my friend Simon in Beaconsfield. Montreal had come off badly from the *Blizzard of 2017* in March, which was followed by heavy rainfall in April that was only three millimetres from setting an all-time record. The rain we experienced that morning only served to top up the perilously high St Lawrence River and the waterfront bike path we were following was frequently diverted due to flooding.

I met Simon in the late 1990's, when I was studying Applied Psychology at Cardiff University. Simon was my tutor, and the university's expert in all things human factors, which is sometimes called ergonomics and is concerned with how we humans best interact with tools and equipment to optimise their safety and performance. This fascinated me. Simon and I bonded over an academic appreciation of human factors and more lowbrow pastimes involving beer and rugby. In 2005, Simon moved to Canada and set up his own c3 Human Factors company. When he discovered our intended route was going past his house, he said it would be churlish not to catch up over a few beers and offered us a place to stay with his family.

Simon's house was easy to find; he'd bravely erected a St George flag in his garden that morning, as a honing beacon. After drying off, we had a late lunch in one of the few riverside pubs which hadn't been temporarily closed due to the weather. Simon told us how he'd embraced Canadian life, developing a passion for ice hockey, skiing and fat bikes; the latter being essentially mountain bikes bedecked in four-inch-plus tyres that can handle deep mud, sand and snow. Fat bikes are becoming popular in the UK but are something of a novelty. However, in Canada, the ubiquitous snow in its harsh winters make them essential for year-round bicycling. We had a relaxing couple of days catching up and then headed for the capital, Ottawa, in much more agreeable weather.

Like Bern in Switzerland or Canberra in Australia, Ottawa is an unlikely capital city. It's not especially well known internationally, it's only the sixth biggest city in Canada population-wise and it isn't even the capital of its province, Ontario (that particular honour is held by Toronto). We joined a tour of the Canadian Parliament building and discovered that Ottawa is only the capital for historic reasons. In 1857 Britain's Queen Victoria decided it should be the capital as it was conveniently located between the important cities of Toronto and Montreal, and it was further from the US border, making it safer from attack.

We spent a few nights in Ottawa with Dutch hosts Stevan and Marieke. They were both experienced cycle tourers and it was great to share experiences of some of the areas that we'd all ridden. We also had the privilege of being invited to a barbeque hosted by Stevan's sister and husband. They too had cycled the West Coast, albeit in much better weather than us.

Ottawa was significant for us as it meant we'd returned to Ontario, the province where we started our tour eleven months earlier. We were close to completing our loop around North America but time was on our side and we decided to make a few more detours in the final week of our mainland adventure.

We headed south from Ottawa until we were reunited with the St Lawrence River. We hadn't realised how close we were to New York State until we saw a sign that simply said Bridge to USA. For a few minutes we contemplated popping over the Seaway Skyway Bridge and riding on the New York side until we reached the Thousand Islands. Sense prevailed when we recalled the trauma of our visas, and we instead continued along the Canadian side of the river to the first capital of Canada, Kingston.

Back when we were in Florida one of our Instagram followers, Dan, had made contact seeking advice about taking a year off and riding around North America with his family. Dan wasn't a novice, having previously spent three weeks riding around Ontario with his wife and three young children. However, he was interested in the challenges a longer-term trip might present and we had a few video calls with Dan to share our experiences. Dan lived in Kingston and, knowing we'd be passing through, offered us a place to stay.

Dan wasn't there when we arrived but, like so many other generous people we'd encountered, left us with details of how to let ourselves in. Whilst we waited for Dan to arrive the next day, we rode to the local mall to meet my friend Andy from university. Andy and I chaired the Motorsport Society together for two years, and we also ran the Inter University Karting Championship, racing teams from twenty-five universities on circuits around the country. This was a low budget affair when we organised it, but has since gone on to become the British Universities Karting Championship with over ninety universities participating. Andy carried his enthusiasm for motorsport into his professional career and has spent the last twenty years leading a jet-set lifestyle as a communications expert in

Formula 1 and MotoGP: the contacts list in Andy's mobile phone would make any fan of motorsport weak at the knees.

We timed our arrival into Kingston just after the 2017 Canadian Grand Prix in Montreal, where Andy was working with Nico Hülkenberg at Team Renault. Andy had decided to take a few days off from his hectic schedule and take his classic Ferrari 328 GTS for a spin around North America. It was rather surreal meeting Andy in the car park of a shopping mall; us on our clunking pushbikes and Andy in his eighties dream car. Nonetheless it was fun to meet in such circumstances and we wished Andy farewell as he tore off looking like an English *Magnum PI*.

When Dan and his family returned home we only managed a quick chat. He was busy juggling his children and his business, but despite this insisted on giving us some invaluable tips for our next stop, Prince Edward County (PEC). On Dan's recommendation we caught the ferry from Adolphustown to Glenora.

PEC is famous for its vineyards, shoreline and the Sandbanks Provincial Park. We were fortunate enough to enjoy all three, having a great day riding around the island and then visiting a few vineyards to sample the goods. After almost a year travelling by bike, I had grown comfortable doing everyday things in my lycra. Ties had stopped wearing cycle clothing early in the trip, favoring the modesty and form of 'normal clothes'. She often had to remind me to contain my own modesty in certain places as I'd become completely immune to the stares and gasps of unsuspecting civilians. The high-end vineyards that I brazenly walked into on PEC were one such occasion. The clientele was generally elderly and draped in floaty linen. I was unaware, but my skimpy lycra was apparently a talking point and was putting everyone off their plonk.

As we pulled into the last vineyard, Harwood Estate, we were told it was unfortunately closing. There seemed to be a few people around so I initially thought this was a ruse to ensure ruffians in lycra didn't lower the tone. However, one of the group, Theshan, asked what we were doing and started firing a barrage of questions at us. Eventually Theshan, who it transpired managed the vineyard, offered us a seat and a glass of wine while more of the group became interested in our antics. After a few more glasses Theshan explained that he and his cousin Ram were laying on a barbeque that night for friends and there'd be plenty of extra food if we wanted to join. Further, we were also welcome to stay the night. We had planned to camp alongside the canal separating PEC from the mainland, so Theshan's offer of staying in a vineyard was quite the upgrade and warmly received. Excited by our new accommodation and dinner plans we had an indulgent night feasting on barbequed meat and imbibing world class wine.

We left PEC in glorious sunshine but within hours there was an almighty thunderstorm. It was grim but we'd been through much worse and rode on wrapped up in all our wet-weather gear. As we rode through the small community of Grafton, we could hear the unnerving sound of a police car slowly trailing us with its sirens blaring. We pulled over, assuming the officer needed to overtake us. Instead, the police officer pulled alongside, wound her window down and bluntly asked what on earth we were doing riding our bikes in such hostile weather. We explained that we'd ridden through vexatious conditions before, and that we were fine. The police officer handed Ties a small cellophaned package and made us promise to wear it. Ties opened it and discovered it was a hi-viz jacket emblazoned with OPP POLICE (OPP is Ontario Provincial Police). I put the hi-viz on – I always cycled behind Ties because I found it impossible to pace myself otherwise – and we enjoyed an afternoon of wide undertaking from drivers that clearly thought we were OPP.

Our last two days in Ontario were spent on the scenic but patchy Waterfront Trail bike path. The trail was a combination of quiet roads, dedicated bikeways and shared boardwalks. It was incredibly scenic but had a frustrating propensity to end suddenly, meaning we were constantly returning to the highway.

OPP Officer Fieldsend

We punctuated the ride with an overnight stay in Ajax. The town does not, as I had wrongly assumed, have long rooted associations with the Netherlands, where the famous Dutch football club Ajax are based in Amsterdam. The town is actually quite new, only established in 1941. Prior to that the area was open farmland and when the Canadians became embroiled in WW2 the government expropriated some of the farms and built an enormous munitions factory in what is now Ajax. The Canadians were firm allies of the UK throughout WW2 and decided to name the new town after the British Royal Navy warship *HMS Ajax*, which had played an instrumental role in Britain's first significant naval victory of WW2: the Battle of the

River Plate, in 1939. The battle took place in Uruguayan waters, where Germany's *Admiral Graf Spee* warship had been ordered to position itself, to disrupt merchant sea lanes once WW2 started. The British sent *HMS Ajax*, along with *HMS Achilles* and *HMS Exeter* to search for the *Admiral Graf Spee.* The British squadron found the German vessel and, after inflicting some damage, shadowed it into the port of Uruguay's capital city Montevideo.

Our hosts in Ajax were Jim and Susie. They were somewhat bewildered by the size of our trip and couldn't believe that we were one day and less than forty miles away from completing our continental loop. We were more interested in finding out about their amazing sustainable lifestyle, which included solar panels so efficient that they were selling energy back to the grid and an extraordinary selection of delicious homemade wines, which they were happy to share.

Fuelled on by a hearty homecooked breakfast, we left Ajax the next day brimming with excitement to be arriving back in Toronto. The trip was far from over – we still had Newfoundland to look forward to – and we were excited to be catching up with Greg and Michelle, who'd had a second child since we'd last seen them a year before. We were also excited to be having a week off the bikes and out of the tent. We thoroughly enjoyed the cycling but the daily routine of unloading the bikes, setting up for one night and then ensuring everything was back in its rightful place before loading the bikes again, was starting to wear thin.

Although we loved the daily adventure and unexpected liaisons with friendly strangers, we were ironically starting to crave a bit of stability. A week in Toronto seemed like the perfect antidote before we headed to our tenth and final Canadian province, which we had no illusions would be the wildest of the whole trip.

CHAPTER NINETEEN

SCREECHED-IN

18th June – 15th July 2017
Distance ridden this chapter: 520 miles / 837 kilometres
Total distance ridden: 14,320 miles / 23,046 kilometres

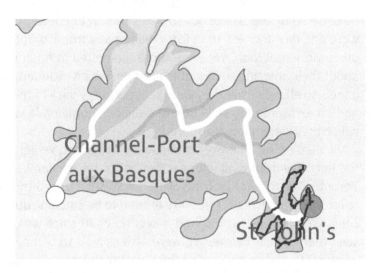

Canadian Prime Minister Justin Trudeau was taking part in a Q&A at Toronto University's Rotman School of Management during our second stay in the city, and we couldn't miss the opportunity to see the Disney Prince (Ties' description, not mine) in the flesh. The session started with reference to the New York Times' recent criticism of what they called the Prime Minister's 'doughnut strategy' to relations with Trump. The implication was that the Prime Minister liaised only with the President's outer circle rather than engaging directly with the central White House. It was entertaining to watch Trudeau bat away the assertions and there was a nervous tension in the room when another journalist dared raise the cakey metaphor later in the debate. I've been accused of projecting an icy stare on occasion, but Trudeau's nonverbal response to that hack beat me hands down.

Our week in Toronto was just what the doctor ordered. Another holiday within a holiday, catching up with friends and languidly exploring more of the city's varied urban environment. We had one small task to accomplish in Toronto; to acquire some Indian wedding outfits for Matt and Dipti's big day. This was easy in the world's most diverse city and it didn't take long in Toronto's Little India to find a helpful shop that decked Ties out in an opulent Saree, and selected a suitable matching Kurta for me. It was surreal to be wedding outfit shopping before we finished our ride, but we had a suspicion it would be easier to find what we needed in Toronto than Newfoundland.

Niagara Falls is only a short drive from Toronto and we took the opportunity of hopping on a Megabus to visit the famous natural phenomenon. The town of Niagara Falls is tragically tacky and there are all manner of expensive excursions to entice those on a healthy budget. Fortunately the waterfall itself is easily viewable from the roadside, of Niagara Parkway, for free and is undeniably majestic. The six million cubic feet of water that gushes over its 160 foot drop every minute is an extraordinary sight and one we're glad we travelled to see.

Whilst at Niagara Falls we also had the opportunity to visit Beverley and Alastair, who we'd stayed with in Fort Myers, Florida. The couple had returned to their home in Jarvis for the summer and we had a great evening catching up on the past five months. Alastair took us out on his boat on Lake Erie and it was a lovely surprise to be joined by Iwona and Irek whose lakeside cabin we'd stayed in on the fourth day of our trip, and who had introduced us to Beverley and Alastair almost a year ago.

For our last night in Toronto, Greg and Michelle had organised a barbeque to send us off in style. We were joined by Greg's parents Lynn and Hank, who had hosted us in New Smyrna Beach (Florida) and Barb, who had flagged us down on the highway almost a year ago and invited us to stay at her house in Thornbury, Ontario. It was wonderful to be reunited with such friendly people who'd opened their doors to us and retained an interest in our exploits months later.

Well-rested, we said our goodbyes to Greg and Michelle the next day and took the three-hour flight to St. John's, Newfoundland, amazed to discover the flight was only two hours shorter than the onward journey would be to London. Newfoundland is a big place – it's the worlds sixteenth largest island – so we had no delusions about riding the whole thing in the three weeks we had available. Instead we decided to ride the 500 mile perimeter of the Avalon Peninsula, named after the legendary island where King Arthur is said to be buried at the foot of Glastonbury Tor, Somerset.

Our first task was to reassemble the bikes. Greg and Michelle had stored our cumbersome bike bags for almost a year in their garage and we were fortunate to find a Warmshowers host in St. John's, called Joy, that agreed to do the same for our tour of the Avalon Peninsula. It was late June 2017 by this point and St. John's was experiencing what we were assured was a rare heatwave. The temperature was tickling twenty-five degrees Celsius – approaching an all-time June record – and it was sweaty work building the bikes in the suntrap that was Joy's garden.

Joy is a fascinating person with a wealth of stories that are reflected in the wonderful character of her home. She has worked as a consultant in environmental economics and policy since 1988 and her passion for sustainability is evident in the way she lives. In 2004 Joy left the comfort of her home and her worldly possessions to live in her Volkswagen Vanagon, Matilda. As well as taking Matilda all over the US and Canada, she travelled to South Africa, Mozambique, Egypt, Vietnam, Malawi and Mongolia. We were captivated by Joy's experiences and insights, and were delighted that she agreed to host us again at the end of our Avalon Peninsula tour.

The fine weather continued as we headed to the stunning

coastal town of Portugal Cove, thought to be named after the Portuguese fishermen that resided there in the seventeenth century. The town will be forever remembered in seafaring folklore as the location where the first evidence of giant squid was brought ashore in 1873. Three fishermen had been attacked by what they thought was a 'Kraken' and managed to lop off two of the beasts' tentacles. They fed one to their dog and sold the other to Reverend Moses Harvey, an avid collector of biological curiosities, for $10. Word soon spread amongst local fishermen and months later a whole specimen was caught in a fisherman's net off nearby Logy Bay. Rev Harvey took a photo of the 'Kraken', which came to be known as 'The Problem of the Giant Squid', referring to the fact the sea creature was now undeniably real.

Scanning for Kraken at Portugal Cove

We made the mistake of spending the night a few miles away, wild camping on the grassy verge above Topsail Beach. It was a beautiful setting overlooking a vast golden bay, but it seemed like every teenager in Newfoundland had decided to party on the beach that night and we hardly got a wink of sleep. I also left with a sore noggin as some rascals had decided it would be fun to throw rocks at our tent, and one of the projectiles caught me square on the forehead. It was painful and at the time I was livid. On reflection, it was extraordinary that this was the only time on the entire trip that we'd experienced any hostility. Further, it was obviously a moronic, alcohol-fuelled deed, rather than a call to arms against cycle-tourers.

We followed the coast the next day, ogling the stunning houses which lined the bays and harbours along the turquoise coast. Our day ended at Clarkes Beach where we'd arranged to stay with Warmshowers hosts Harold and Whit for two nights. Harold found out that we hadn't been 'screeched-in' and insisted we observe the tradition. Being screeched-in is a rite of passage for any 'Come-From-Aways'; anyone visiting or living in Newfoundland who isn't born there. Harold kicked off the ceremony by asking us "Is ye a Screecher?", to which we replied "'Deed I is me old cock, and long may your big jib draw."

True to tradition we then ate some bologna ('Newfoundland steak'), kissed a cod and downed a shot of screech (rum). We are both now honorary Newfoundlanders.

The odd tradition has been forever immortalised in the musical *Come from Away*, which depicts the extraordinary hospitality shown by Newfoundlanders in the wake of the 9/11 attacks. US airspace was closed soon after the towers collapsed forcing thirty-eight aircraft to land at the remote Gander airport in Newfoundland. The 7,000 passengers doubled the population of the town but the residents threw open their doors giving the passengers shelter until onward travel could be arranged. In the musical, the song 'Screech In' tells how the local bar offered to calm down any nervous passengers with the same ceremony that we experienced for fun. We would go on to discover that the extraordinary hospitality of the residents in the musical is still alive and well, years after the tragic events of 2001.

We headed north, hopeful we'd see some of the icebergs that Harold and Whit assured us were floating off the northern coast of the Avalon Peninsula. The morning's ride was another hilly affair, rewarding us with empty roads and rugged coastline. As we passed the aptly named Small Point, we couldn't believe our eyes. There was an enormous iceberg, shaped almost like a half pipe skateboarding ramp, floating just off the shore. We rode down to Mulleys Cove and spent the next two hours staring at the incredible sight, whilst being wafted by an icy breeze that we imagined emanated from the iceberg.

As we were leaving there was an almighty noise like a thunderclap. Initially we were worried the dark skies were turning into a storm but when we glanced back at the iceberg, we realised a huge chasm had divided the glacial mass and it was parting into two smaller icebergs. We couldn't believe that not only had we been fortunate enough to see such a mighty iceberg but we'd also witnessed the ancient structure splitting. A truly unforgettable experience that is as vivid now as it was in 2017.

Newfoundland has many creatively named areas that never fail to raise a smile or an eyebrow. We rode through a particularly entertaining stretch of coastal villages called Heart's Content, Heart's Desire and Heart's Delight. No one really knows where the names come from, but that doesn't stop locals speculating about heart-shaped coves, heartstring pulling scenery and heart stealing mermaids.

Just south of the 'Hearts' is the community of Dildo, which is genuinely opposite Spread Eagle Bay. Speculation is rife about the origins of the blush-inducing names, but the favourite explanation is that cheeky Captain Cook intentionally gave the areas humorous names when he mapped Newfoundland in the eighteenth century. In recent times, petitions have failed to change the name of the town and in 2019 comedian Jimmy Kimmel brought international attention to Dildo when he commissioned Hollywood-style lettering to be erected in the hill overlooking Dildo Cove. Rather than take offence to the stunt, local businesses were delighted at the additional revenue the sign brought in and made Jimmy the first honorary mayor of Dildo. Jimmy accepted the title saying "Thank you my brothers and sisters in Dildo, you've been very kind. I'm proud to be your mayor and I will come to visit my fellow Dildodians very soon."

The majestic iceberg that entertained us for hours

The friendliness of Newfoundlanders is such that we frequently found ourselves discussing our intended route with strangers. Every day I'd take out my paper map and talk several people through the itinerary. As I traced my finger along the roads one stretch never failed to induce a deep intake of breath; the Cape Shore on the southwestern section of the Avalon Peninsula. As we left our grubby hotel room in the curiously named town of Placentia, we soon discovered that warnings about the hills were not exaggerated. The worst climb of all was the ascent from Great Barasway: the relentless climb seemed to go on forever, without any opportunities to rest and no sight of the top. I crunched my way to the top and waited for Ties. Eventually a car pulled alongside and told me that someone who looked like they might be with me was pushing their bike and looked to be in pain. I propped my bike up against a pine tree and walked down the hill, finding Ties in agony. The climb was too steep for her to ride and she'd cricked her back trying to push her bike up the hill. I rode Ties' bike up the remainder of the hill and she had no choice but to walk up in considerable pain. We were only thirteen miles into our day, which promised lots more long, steep climbs. Our only option was for me to keep riding both bikes up the remaining hills, with Ties walking up and jumping back on her bike when the gradient was more agreeable. It was undoubtedly the hardest thirty miles of the trip and we ended the day as soon as we could at the small town of Saint Bride's.

After such a difficult day we were not feeling particularly fussy about where we spent the night. We spotted a small school in Saint Bride's and asked the caretaker if we could pitch our tent on the side of the building. He was reluctant to authorise our request but a group of children that were playing nearby came to the rescue. They told us their mum was the head teacher and they said they would go and ask if it was okay. Minutes later the children scurried back and excitedly told us we had their mum's permission. The caretaker told us he'd leave the school open that night, which meant we had access to drinking water, toilets and power to charge up all our gadgets.

The climb from Great Barasway had given Ties intense back

pain the following day. This wasn't helped by the persistence of the Cape Shore's hilly terrain, or the fact we ended that day by sleeping on the plot of an abandoned grocery store in Mount Carmel. It wasn't glamorous but it had a level surface and we were beginning to understand that it wasn't just the gradient of Cape Shore that made it such a tough place to explore. Amenities were few and far between and while we desperately wanted a shower, we didn't have the energy to keep hunting for better options.

***

Countless people had told us about the annual capelin roll, a strange natural phenomenon that sees millions of the small capelin fish barrelling onto shore to spawn. The event has been witnessed for centuries and folklore holds that summer cannot start in Newfoundland until the capelin are rolling. There are so many of the fish that locals walk into the surf with buckets and simply scoop them out of the ocean. The main excitement surrounding capelin is not the tasty dishes they make for humans, but their appeal to a much larger mammal, the humpback whale. The whales are canny to the habits of the capelin and funnel the fish into coves ready for suppertime. Like the petite fish, whales are also creatures of habit and we were told that if we arrive at Saint Vincent's Beach at precisely 7pm we'd witness something very special.

Saint Vincent's was only forty miles from our functional overnight spot in Mount Carmel. Having spent two nights roughing it we decided to have an early start and stay in the relative luxury of the B&B in Saint Vincent's. Everything went to plan. The weather played ball, the B&B had one room available and the owner confirmed that you can set your watch by the behemoth whales. We washed off three days of filth and made our way to the beach. Sure enough at 7pm the first whales arrived, waving their colossal fins at us and slapping their tails firmly on the ocean surface. Within minutes the cove was full of whales, putting on a show that looked like it was for our benefit rather than theirs. I was struggling to count them but a local told us it was a good night and there were at least thirty, all within clear sight of shore. No binoculars needed.

We thought our incredible whale watching experience at Saint Vincent's would be a one-off but we'd soon discover that whales in Newfoundland are like alligators in Florida; they take a while to spot, but once you see one you can't stop seeing them. We would go on to have many more sightings, all from the shore and all as spectacular as the first.

Whale watching on Saint Vincent's Beach

Our dalliance with luxury at the Saint Vincent's B&B was short lived and the following night we were back to wild camping in the picnic area of the tiny town of Portugal Cove South. I had heard that there were some noteworthy fossils at the nearby Mistaken Point Ecological Reserve. So significant are these fossils, the 'oldest fossils of complex multicellular life found anywhere on Earth', that in 2016 the area had been designated an UNESCO World Heritage Site. We arrived too late for the days' tour but were told we could attend the following day and that camping in the town's picnic area "should be fine."

Ties was keen to rest her back and chilled out in the tent as I became a fossil-hunter for the day, adroitly guided by local palaeontologists. The tour was fascinating and it wasn't hard to spot the large fossils that carpeted the flat rocks along the coast.

After two nights of camping in the town we were feeling a bit self-conscious and were on the road by 7am the next morning. Within minutes the heavens opened and we were completely drenched with nowhere to shelter in the barren mass of the Avalon Wilderness Reserve. Twenty-five sodden miles later we saw a sign swinging frantically in the wind, at the small fishing village of Renews, which included the magic words *Coffee* and *WiFi*. We pulled over and a man came to the door ushering us into the building in a thick Irish accent. The inside of the building was far grander than its ramshackle exterior let on. There was art and crafts everywhere, as well as expensive looking handmade clothes and a giant loom in the middle of the room.

We were hurried to a table and before we even sat down coffees were presented, accompanied by delicious looking cake. The man at the door was John. He and his wife Rita proceeded to tell us all about the art gallery cum café they'd recently created from the building that Rita's father had used for years as a net loft. The couple told us that they also held events in the gallery and their son-in-law Graham, a renowned Celtic accordion player, was holding an intimate concert there that evening. Within no time we were invited to the concert and soon after that the couple left to attend a funeral, leaving us in charge of the gallery.

Throughout the day we were introduced to more and more of John and Rita's family and by mid-afternoon had been taken under the wing of Aunty Beth, whose house we were told we'd be staying at. The concert was a truly unique experience, with

Graham playing some haunting music inspired by Newfoundland's fishing heritage and harsh climate. When the public performance finished the whole family piled back to John and Rita's where Graham gave an even more intimate encore, interposed by John enthusiastically reciting ghost stories and tales of yore.

***

We slowed things down in the last week of the trip. We'd already ridden 2,000 miles further than anticipated and we decided to have some short days for the final stretch to St. John's. All that stood between us and Newfoundland's capital was seventy-five miles of coastline. Delightfully, it turned out to be the most scenic seventy-five miles of the whole trip. That one small strip of coast was like riding through the *Greatest Hits* of North American Scenery. The water was crystal clear, alternating between turquoise rivers and inlets, to swimming-pool-blue bays and intense dark navy-blue coves. The roads were perfectly tarmacked and gave contrasting views of Christmas tree conifers and rocky outcrops of all shapes and sizes. When the quiet highway left the rugged shoreline it snaked through postcard-perfect ponds dotted with enviable log cabins and verdant islands. One enterprising local had unashamedly called his eatery Bernard Kavanagh's Million Dollar View Restaurant, which seemed like a fitting moniker. And of course we saw countless humpback whales, all dining on capelin, some pugnaciously ignoring the 7pm rule.

The *crème de la crème* was not Bernard Kavanagh's but La Manche Provincial Park. It can't claim to be the best named Provincial Park in Newfoundland – which must surely be a battle royale between Blow Me Down, Butter Pot, Dildo Run and J.T. Cheeseman – but for our money it gave us the single best view of the entire year in North America. The spellbinding sight of the awe-inspiring wooden suspension bridge, framed by the verdant boreal forests, was accompanied by the aroma of pine and tinkling of the river. A feast for all the senses.

It seemed fitting that La Manche was our last camping spot and that our final stretch to St. John's was again 'Dartmoor hilly'. We took the final miles slowly and spent a few days in the capital hanging out with Marie and Douglas, new friends that we made at the concert in Renews. The couple live in Calgary but have a holiday home in St. John's. They took us down to the impossibly idyllic Quidi Vidi Harbour, where we enjoyed a tasty pint of Iceberg Beer at the waterfront taproom of the Quidi Vidi Brewing Company. Another curiously named Newfoundland community whose etymology remains a mystery.

The wooden suspension bridge at La Manche Provincial Park; our favourite view of the whole trip

Quench sated, Marie and Douglas drove us to the historic landmark Signal Hill. In 1901 budding Italian inventor Guglielmo Marconi made history by sending the world's first transatlantic radio message from Poldhu in Cornwall to the abandoned fever and diphtheria hospital on Signal Hill in St. John's. Nowadays the hospital is long gone but Cabot Tower remains as both a legacy to the historic event and an exceptional lookout point. When we there we witnessed the heart-in-mouth scene of an enormous humpback whale approaching a lone kayaker, diving under the vulnerable watercraft and emerging just far enough ahead to not topple the panic stricken pilot. There was a well-choreographed collective intake of breath from all those on Cabot Tower, followed by a communal exhale and round of applause when the whale reappeared at a safe distance. The kayaker looked mightily relieved too.

***

In his long and illustrious career as a taxi driver, the chap who took us from St. John's to St. John's Airport had claimed never to have seen bike bags before. He couldn't understand how bikes fitted into the strange shaped luggage and had an even harder time comprehending the 520 miles we'd ridden around the Avalon Peninsula. It took the entire drive from the city centre to the airport to convince him that we'd done this, by which point we didn't have the time or the inclination to point out that it was a small part of our 14,320-mile bicycle tour of North America.

We made our way through security and headed straight for the Airport's Tim Horton concession to consume one last box of *Timbits*; the flavoured fried dough balls that had lifted our spirits so many times on our trip and gave us one last taste of Canada before we flew back to reality.

EPILOGUE

ANTS IN THE PANTS

Why now? Why write this book four years after we finished our trip? The short answer is Covid-19. As I write this in October 2021 we should be nearing the end of our second cycling adventure. Sadly we've had to postpone this trip indefinitely, pending the worldwide rollout of vaccines, the return of some semblance of normalcy and the acquisition of funds to afford another long bike tour. The more accurate response to the question is less brusque and decidedly more hopeful.

When we were riding around North America we were asked to recount our stories numerous times. It was amazing to us how excited complete strangers were by our adventure and the conversation would normally end with 'you must write a book'. To which we would always respond 'we are definitely going to'. When we returned to the UK the reality was somewhat different to the image I had of sitting at a desk, gleefully bashing out a tome worthy of publication.

We had to find somewhere to live, we had to find work, we had endless reunions all over the country and frankly we grew tired of retelling the same stories over and over again. The last thing we wanted to do was write a book. In early 2018, less than two years after returning from North America, we were already plotting our next adventure and within months we had a feasible plan.

We intended to ride through Southeast Asia after Christmas – year to be determined – then fly to Japan in late March, to see the cherry blossom. After riding to Hokkaido Island, in Northern Japan, we planned to head back to London, and then spend the summer exploring Scandinavia in our campervan.

***

Epilogue

We were living in Falmouth, Cornwall, when we hatched this plan. I loved Falmouth. The cycling and scenery are exceptional and to my biased British eyes, the best in the world. Ties was struggling with the isolation and was not a fan of the hilly terrain. She missed cycling and I had to admit that the monstrous car journeys required for both work and leisure made the reality of living in Cornwall less than ideal.

Covid-19 kiboshed our plans to start our next adventure in January 2021. Unsure when international adventuring would again be feasible, we decided to buck the national trend of moving from city to countryside and move to Bristol.

On a recommendation from my dad's colleague, Matt, I read Ian Walker's book, *Endless Perfect Circles*. The book tells the inspiring story of how Ian transitioned from non-cyclist to ultra-distance World Record holder, in a few short years. I have my own ultra-distance aspirations and found Ian's storytelling addictive and compelling.

We were enjoying our new life in Bristol, but with every day that passed I couldn't help looking at the schedule I'd prepared – yes, I mapped the whole route out again as I did for North America – and pining after the experiences we were missing out on. Inspired by Ian's book, and a desire to override these painful thoughts with blissful memories, I decided to spend my spare time penning this book and reliving the trip.

I've drawn on a number of resources to write this book, including our Instagram account (*cyclinktheworld*) and some blogs that we posted during our trip. I maintained a simple log on my iPhone's Notes App – every night before I went to sleep I would record the start and end location for the day, the mileage, where we stayed and the cost (if we stayed in paid accommodation) – that proved invaluable for writing this book. I didn't originally do this for the benefit of writing a book. I just love data! Aside from my own resources, I also did a fair bit of Googling for many of the book's historical snippets.

Writing this book has been a truly enjoyable experience. It's not always been easy to do so four years after the fact, but it's been an absolute pleasure to relive so many memorable events and recall some less obvious happenings that nonetheless made the year what it was.

At some point in the future we'll load up the bikes and hit the road again. It will be a very different trip to our North American adventure. For one, we are now vegan, so we'll have a new fuelling challenge to overcome. We also learned numerous lessons from our first cycle tour – number one being that weather is the biggest challenge – which we'll take with us on our next trip.

Our next big ride may not be the *Southeast Asia to Japan to Scandinavia* adventure we planned for 2021, but wherever we go we'll be hopeful that our wheels will bring us the fortune that we experienced over and over again in North America.

ACKNOWLEDGEMENTS

This book would have been very different where it not for the huge amount of people who opened their hearts, minds and houses to us on our trip.

I cannot start this section without thanking Ross Daliday. Ross planted the adventure travel seed in my head and then helped it germinate with persistence and hops-flavoured beverages. Ross also edited this book, patiently and candidly reviewing all the chapters again and again until its novice author finally produced some readable material.

Huge thanks to Sophie Houghton, for goading me into a holiday to Mexico. Had I not followed Sophie's advice I wouldn't have met Ties, and I can guarantee this book would have been much more boring.

I'm extremely grateful to award-winning author Diane Holiday, for providing expert advice on the art of authoring and reviewing this book, and to Racqui Saines for her invaluable proofreading. Thanks to Rob Franklin (of E3D Studio London Ltd) for the excellent cover and route maps, and to Ian Walker for advice and guidance on how to self-publish a book.

Much love to my parents, Margaret and Mally Fieldsend who gifted us an incredible all-expenses-paid holiday on Vancouver Island. Not to mention, provided storage for the belongings we didn't discard before leaving the UK.

Thanks to Tom Donhou for building me such an exquisite riding machine, and providing transatlantic support and guidance throughout the trip.

Although our visa experience was tough, it would have been much tougher if it wasn't for the consistent support of Jon and Jess Easthope. Thanks also for hosting us in Renton and helping Ties recuperate.

Enormous thanks to all the generous soles, friends and strangers alike, that supported us before and during our trip.

With apologies to the many people who helped us but whose name I never even found out, thanks to Aaron Brown, Alex Siegel, Andrew and Pam, Andrew Ford, Andrew Ware, Barb and Chuck Reynolds, Ben Woodard, Beth Chidley, Beverley and Alastair Robertson, BikeBike (Calgary), Brent, Diane and Steve Holiday, Cal (Nova Scotia), Camping d'Amqui, Channing Frampton, Colin White, Colleen Sampson and Bill Gochicoa, Dan McDougall, Devon (Doodlepuffy), Dutch Cycles (Regina), Ed ('Canada's Rohloff expert'), Eddie Xing, Edith and Winston, Elise Roper, Ellie and Diane Baecklund, Eugene and Birdie Wiebe, Fatima Academy, Gabriel Lang, Gareth Neufeld, Ghisleli Ramirez, Ginger Mills, Greg Appleton, Greg Spencer and Michelle Astrug, Harriet Booth, Hector and Marie, Hubbards Beach Campground & Cottages, Ingrid and Mike, Iwona, Irek and Kasia Stepiens, James Bannochie and Emma Ballinger, Jamie Leathem and Patrick Little, Jane Henderson and Dennis Duermeier, Jay Borden (Roulez Cycles), Jennifer Qualls, Jocelyn Rice, John and Rita Chidley, Julian Vera, Kathryn Zimmerman, Katya Hesse, Ken Dillon, Kevin Ryan, Kiefer Blackburn (Flat Fix, Vancouver), Leo Malpica, Les Schevelik, Linda Allsop , Linda Garrett and Joyce Mixon, Lisa Broom, Loki (Vancouver), Lynn and Hank Spencer, Marie Chidley and Douglas Meggison, Mathieu and Léonie, Matt Starling, Megan Woodward, Mining Journal (Marguette), Morton (MacGregor), Neil Pengilley, Nick (Kermode Cycles), Nick Gordon, Ontario Provincial Police, Paul McKenzie , Philippa and Robert Panjer, Rachel Wise and Simon, Ruby (Bleakwood), Rupert and Martina Staudner, Ryan Davies, Sharon Sleeper, Shirley Nadeau, Simon Banbury and Jennifer Landry, Simon Nankivell, Tal, Texas County Sheriff, Theshan Puvi, Tina (Coastline Cyclery), Tommy and Richard, Tyler and Kaci, Uncle Mo and Yves Cadieux.

Warmshowers changed our trip significantly and changed my outlook towards hosting and being hosted. It brought not just shelter and comfort, but new friends and new experiences. Thanks to Warmshowers hosts Alex and Anouk, Alison Martin,

Allison And Jimmy Vukelich, Austin Blessard, Barbara and Jarlath, Barbie Fallon, Benjy and Susie, Bill Vanderwall, Bruce and Vikki White, Bruce Hart, Christine and Alex Bergeron, Christopher Lewis, Cindy Davis, Craig and Dianne Skills, Cyndi Bakir, Dan and Kim Sees, Dave Depp and Kathleen Kenney, Dustin and Meghan Dahn, Fletcher Yancey, Gabriele Marewski, Harold Moore and Whit, Isaac Bonisteel, James Tremblay, Jeffrey and Heidi Ellis, Jerry and Lainie, Jim and Susie Blakelock, John Trahan & Laura Elsenboss, Jordy Carrier, Joy Hecht, Judith, Kees Keizer and Hilary, Keresha and Chris, Kerry and Alyssa Lash, Kevin Motel and Genna, kt Misener and Rick Willing, Larry and Kimberley Hansen, Larry W, Len M, Lenny Greene, Lola, Margit Pirsch and James Polack, Mark and Debbie Gutekunst, Mark, Carol and Christian Jacopec, Mary and Bernie Roycroft, Matt Fraser, Matt Muise, Michael K, Michael Warner, Mike Zysman & Jackie Brady, Nathalie Turgeon, Nick O'Connell, Pam and Tom Frisk, Pascal Walde Ruzzante, Pat Broom, Paul Molyneaux, Randy and Cheri Hays, Richard Humphreville, Rob and Cindy Macpherson, Rob Edgren, Robbie Wood, Stephen Tham, Stuart and Delores Mast, Taylore Becker and Luke, Tom and Nancy Peck, Tyler Carpenter and MaryAlice Misuta, and Wayne Pinkerton.

Most of all, thank you to Ties. Thank you for ditching your life in the tropics for dreary Blighty. Thank you for placing your complete faith in me to undertake such an incredible physical and mental challenge. Despite all the trials and tribulations our trip threw up, thank you for continuing to have faith in me and wanting to keep on adventuring with yours truly.

WORLD BICYCLE RELIEF

All profits from the sale of this book support the work of World Bicycle Relief (WBR):

"World Bicycle Relief mobilises people through The Power of Bicycles. We are committed to helping people conquer the challenge of distance, achieve independence and thrive. In developing countries, millions of people walk for miles each day just to survive. On foot, individuals race against the sun to complete everyday tasks. Distance is a barrier to attending school, receiving healthcare, and delivering goods to market.

World Bicycle Relief delivers specially designed, locally assembled, rugged bicycles for people in need. We've developed an efficient, innovative, and scalable model to empower students, health workers, and entrepreneurs in rural developing regions with life-changing mobility

We send humble gratitude on behalf of the bicycle recipients that this book has helped."

– *World Bicycle Relief*

Why WBR? Indulge me, while I get a little geeky.

The pinnacle of every professional road cyclists' career is riding the Tour de France (TdF), a brutal test of endurance that involves riding over 2,000 miles in three weeks. The winner of the overall event receives the *maillot jaune* (yellow jersey), the best young rider is awarded the white jersey, the polka dot jersey goes to the king of mountains (best climber) and the best sprinter wears the green jersey.

I thought the 2021 TdF was enthralling. The Slovenian whippersnapper Tadej Pogačar took the white, polka dot and yellow jerseys for the second year running. For fun, Pogačar then went and picked up a medal in the Olympic road-race in

Tokyo.

The green jersey went to Mark Cavendish. If the Brothers Grimm were commissioned by the UCI (governing body for world cycling), they wouldn't have been able to write a fairytale like Cav delivered at the 2021 TdF. If you've read this book from front to back – firstly, well done – you'll know I (sort of) took part in the 2007 TdF. So did Cav. That's how long the Manx Missile has been riding professionally. Cav's sprint *palmarès* are astonishing. He dominated the end of flat stages in the TdF for years. Unfortunately, in April 2017 he was diagnosed with the Epstein–Barr virus, which can lead to glandular fever. Cav had been experiencing 'unexplained fatigue' and the illness would go on to impact the rider for two years.

At the ripe old age of thirty-three, Cav's professional career was on the ropes. In 2021, he was recruited by the Belgium professional cycling team Deceuninck–Quick-Step, whose lead sprinter then withdrew from the TdF. The Manx Missile was thrust back into the spotlight and soppy fans like me dared to believe. Boy did he deliver. Cav won four stages and equalled Eddy Merckx's all-time record for most TdF stage wins.

Tadej and Cav's 2021 TdF achievements are record-breaking. However, for me, a little-known Australian rider is *the* story of the race. That rider is Lachlan Morton, who rides for professional cycling team Education First-Nippo. He is more than twelve years my junior, and basically, he's my hero. Lachlan has competed in, and won, numerous ultra-cycling races all over the world. This includes the inaugural 2019 GBDuro, an all-road 1,200 mile race from Land's End to John o' Groats. Lachlan completed the GBDuro in less than five days, beating his nearest competitor by over thirty-eight hours. In June 2020, unable to ride hundreds of miles due to Covid-19 restrictions, Lachlan had a go at everesting: he set a new world record. Then at the 2021 TdF he completed the 'Alt Tour': riding the actual TdF route, solo and unassisted, plus all the transfers from the end of one stage to the start of the next. At one point, Lachlan experienced knee pain and replaced his clipless pedals for flats. Undeterred, he continued the ride in a pair of modified Birkenstock sandals. Not exactly *vêtements de cyclisme habituels*.

Despite riding almost 1,000 additional miles to the official TdF racers, Lachlan arrived at the Champs-Élysées five days ahead of the peloton. My mind boggles trying to conceive how Lachlan rode over 3,400 miles in sixteen days, fully self-supported, including a stretch in Birkenstocks! When the Alt Tour idea was conceived, Lachlan decided it would also be an opportunity to raise money for WBR (at the time of writing, £521,267). After a few days of following the Alt Tour I, like thousands of others, clicked the link to donate to WBR. As I was entering my bank details I had a revelation: rather than donate, why don't I give the profits of this book to WBR. My motivations for writing this book were never money – I wanted to tell our story, inspire others and restore a modicum of faith in humanity – so supporting WBR in this way seemed logical.

ALMANAC

I am a big fan of data and couldn't resist including a summary of the statistics from our trip. I hope someone out there finds this half as interesting as I do!

Distance

- Longest days (91 miles)
    o Swift Current to Maple Creek (Saskatchewan)
    o Rodeo to Columbus (New Mexico)
- Shortest day (13 miles)
    o Greenpoint to Prospect Park South (New York)
- Most days without a rest day (twelve days, covering 646 miles)
    o Portland (Maine) – Riverport (Nova Scotia)
- Most distance without a rest day (683 miles, in eleven days)
    o Rodeo (New Mexico) – Uvalde (Texas)

Mechanicals

- Ties had fourteen punctures and I had seven
- We both changed our tyres just the once
- Ties went through seven chains and one cassette
- I went through one belt, one shift box and one pair of gear shifters
- I made four oil changes to the Rohloff geared hub
- Ties replaced five pairs of disc-brake pads and I replaced two pairs
- Ties replaced one rear gear cable. Remarkably, we didn't replace any other cables, and they're still going strong to this day!

Table 1: Summary of data from each Canadian province and US state

Canadian Province / US state	Dates	Riding days (full)	Rest days	Total days	Average daily distance*	Total distance
Florida (US)	Jan-Feb 2017	27	6	33	56	1,521
California (US)	Oct-Dec 2016	31	8	39	46	1,417
Ontario (Can)	Jul-2016; Jun-2017	19	9	28	53	1,007
British Columbia (Can)	Aug-Oct 2016	24	17	41	39	933
Texas (US)	Dec 2016 - Jan 2017	17	3	20	54	911
Quebec (Can)	May-Jun 2017	11	3	14	63	688
Nova Scotia (Can)	Apr-May 2017	16	3	19	40	634
Newfoundland and Labrador (Can)	Jun-Jul 2017	14	8	22	37	520
Arizona (US)	Dec 2016	10	1	11	51	512
Maine (US)	Apr 2017	10	1	11	46	458
Saskatchewan (Can)	Aug 2016	7	1	8	62	437
Alberta (Can)	Aug 2016	7	2	9	56	389
Oregon (US)	Oct 2016	8	4	12	47	372
North Carolina (US)	Mar 2017	6	1	7	62	370
New Brunswick (Can)	Apr 2017; May 2017	7	0	7	51	355
Louisiana (US)	Jan 2017	6	3	9	59	354
Prince Edward Island (Can)	May 2017	6	2	8	58	347
Manitoba (Can)	Jul-Aug 2016	5	1	6	65	326
Michigan (US)	Jul 2016	5	1	6	64	319

Almanac

Canadian Province / US state	Dates	Riding days (full)	Rest days	Total days	Average daily distance*	Total distance
Virginia (US)	Mar 2017	7	2	9	42	297
South Carolina (US)	Feb-Mar 2017	6	1	7	49	294
Washington (US)	Oct 2016	6	2	8	48	290
New York (US)	Mar-Apr 2017	6	3	9	35	212
Minnesota (US)	Jul 2016	3	0	3	65	196
New Jersey (US)	Mar 2017	4	0	4	45	178
New Mexico (US)	Dec 2016	2	1	3	84	167
Massachusetts (US)	Apr 2017	3	1	4	48	145
Georgia (US)	Feb 2017	3	1	4	47	141
Alabama (US)	Jan 2017	2	0	2	57	114
Wisconsin (US)	Jul 2016	2	1	3	55	109
Mississippi (US)	Jan 2017	2	0	2	47	93
Maryland (US)	Mar 2017	1	0	1	69	69
Rhode Island (US)	Apr 2017	1	0	1	42	42
Delaware (US)	Mar 2017	1	0	1	38	38
Connecticut (US)	Apr 2017	1	0	1	27	27
New Hampshire (US)	Apr 2017	0	0	0	20	20
District of Columbia (US)**	Mar 2017	0	2	2	18	18
Canada		116	46	162	49	5,636
US		170	42	212	51	8,684
Total		286	88	374	50	14,320

*All distances in miles
**I know DC isn't *really* a state!

Table 2: Percentage of nights spent by each accommodation type

Canadian Province / US state	Airbnb	Camp-ground	Friends	Motel	Strangers	Warm-showers	Wild camping
Florida (US)	6.1%	18.2%	15.2%	12.1%	18.2%	24.2%	6.1%
California (US)	15.4%	41.0%	0%	7.7%	2.6%	25.6%	7.7%
Ontario (Can)	0%	35.7%	25%	7.1%	17.9%	14.3%	0%
British Columbia (Can)	36.6%	53.7%	2.4%	2.4%	4.9%	0%	0%
Texas (US)	10%	30%	30%	20%	5%	5%	0%
Quebec (Can)	21.4%	21.4%	14.3%	21.4%	14.3%	7.1%	0%
Nova Scotia (Can)	0%	5.3%	0%	31.6%	0%	36.8%	26.3%
Newfoundland and Labrador (Can)	9.1%	13.6%	0%	13.6%	4.5%	36.4%	22.7%
Arizona (US)	0%	9.1%	0%	9.1%	9.1%	63.6%	9.1%
Maine (US)	36.4%	0%	0%	0%	9.1%	54.5%	0%
Saskatchewan (Can)	0%	100%	0%	0%	0%	0%	0%
Alberta (Can)	33.3%	66.7%	0%	0%	0%	0%	0%
Oregon (US)	41.7%	50%	0%	8.3%	0%	0%	0%
North Carolina (US)	0%	42.9%	0%	0%	0%	57.1%	0%
New Brunswick (Can)	0%	14.3%	0%	28.6%	14.3%	42.9%	0%
Louisiana (US)	0%	0%	0%	33.3%	0%	66.7%	0%
Prince Edward Island (Can)	0%	12.5%	0%	25.0%	62.5%	0%	0%
Manitoba (Can)	33.3%	50.0%	0%	16.7%	0%	0%	0%
Michigan (US)	0%	83.3%	0%	16.7%	0%	0%	0%
Virginia (US)	11.1%	0%	0%	11.1%	0%	77.8%	0%
South Carolina (US)	14.3%	14.3%	0%	14.3%	0%	28.6%	28.6%
Washington (US)	0%	62.5%	37.5%	0%	0%	0%	0%
New York (US)	0%	0%	0%	0%	33.3%	66.7%	0%
Minnesota (US)	0%	100%	0%	0%	0%	0%	0%

Almanac

Canadian Province / US state	Airbnb	Camp-ground	Friends	Motel	Strangers	Warm-showers	Wild camping
New Jersey (US)	0%	0%	0%	0%	25%	75%	0%
New Mexico (US)	0%	100%	0%	0%	0%	0%	0%
Massachusetts (US)	0%	0%	0%	0%	0%	100%	0%
Georgia (US)	0%	0%	0%	25%	0%	50%	25%
Alabama (US)	0%	0%	50%	0%	0%	50%	0%
Wisconsin (US)	0%	100%	0%	0%	0%	0%	0%
Mississippi (US)	0%	100%	0%	0%	0%	0%	0%
Maryland (US)	0%	0%	0%	0%	0%	100%	0%
Rhode Island (US)	0%	0%	0%	100%	0%	0%	0%
Delaware (US)	0%	0%	0%	0%	0%	100%	0%
Connecticut (US)	0%	0%	0%	0%	0%	100%	0%
New Hampshire (US)	-	-	-	-	-	-	-
District of Columbia (US)	0%	0%	0%	0%	100%	0%	0%
Canada	15.4%	35.8%	6.2%	12.3%	9.9%	14.2%	6.2%
US	9.9%	28.3%	7.1%	9.9%	7.5%	33%	4.2%
Total	12.3%	31.6%	6.7%	11%	8.6%	24.9%	5.1%

ABOUT THE AUTHOR

Chris is a two-wheeled enthusiast who spends as much time as possible riding his bike long distances around the UK. He developed a fondness for mountain biking in the early 1990's, before trading knobbly tyres for skinny rubber and becoming a dedicated roadie.

Chris loves a challenge and has raised thousands of pounds for charity by completing a stage of the Tour de France on a fixed gear bicycle, and riding 330 miles from London to Amsterdam in less than forty-eight hours. In 2016-2017, along with his girlfriend Ties, he went further afield, riding 14,320 miles (23,046 kilometres) around all ten Canadian provinces and twenty-six US states.

When not riding his bike, Chris works as an independent consultant, advising organisations how to optimise their programme management and build data maturity.

Chris and Ties live in Bristol, southwest England. The couple have grand plans for more bicycle-based adventures.

Chris and Ties' Instagram profile (*cyclinktheworld*) has colour copies of all photos in this book, along with hundreds of other photos from their adventure.

Made in the USA
Monee, IL
03 October 2023

43896708R00152